Springer Japan KK

K. Okita (Ed.)

Growth, Proliferation, and Apoptosis in Hepatocytes

With 53 Figures

 Springer

Kɪᴡᴀᴍᴜ Oᴋɪᴛᴀ, M.D., Ph.D.
Professor and Chairman
Department of Molecular Science & Applied Medicine
Gastroenterology & Hepatology
Yamaguchi University School of Medicine
1-1-1 Minamikogushi, Ube, Yamaguchi 755-8505, Japan

ISBN 978-4-431-68001-7

Library of Congress Cataloging-in-Publication Data
Growth, proliferation, and apoptosis in hepatocytes / K. Okita. ed.
    p. ; cm.
Includes bibliographical references and index.
    ISBN 978-4-431-68001-7   ISBN 978-4-431-67887-8 (eBook)
    DOI 10.1007/978-4-431-67887-8
    1. Liver—Pathophysiology—Congresses. 2. Liver cells—Congresses. 3.
Apoptosis—Congresses. I. Okita, Kiwamu. II. Yamaguchi Symposium on Liver Disease
(12th : 2000)
    [DNLM: 1. Hepatocytes—cytology—Congresses. 2. Apoptosis—Congresses. 3. Cell
Physiology—Congresses. 4. Liver Diseases—physiopathology—Congresses. 5. Liver
Regeneration—physiology—Congresses. WI 700 G884 2002]
    RC846.9 .G76 2002
    616.3'6207—dc21
                                                                2001049885

This symposium was supported by funds from the Viral Hepatitis Research Foundation of Japan.

© Springer Japan 2002
Originally published by Springer-Verlag Tokyo Berlin Heidelberg New York in 2002
Softcover reprint of the hardcover 1st edition 2002

SPIN: 10853578

Preface

Since the 1st Yamaguchi Symposium on Liver Diseases in 1989, this series of symposia has provided opportunities for exchanges of information on the topic between leading Japanese hepatologists and internationally renowned scientists. Somewhat unusually for meetings held in Japan, the official language of the symposium is English. The proceedings of these symposia are published under the title *Frontiers in Hepatology* and distributed worldwide.

The 12th symposium was held on December 9 and 10, 2000, at the ANA Hotel, Ube, Japan. The theme selected by the Organizing Committee was "Growth, Proliferation, and Apoptosis in Hepatocytes," each of which is important in the understanding of the pathophysiology of intractable liver disease. Nine Japanese hepatologists were invited to give presentations, as was leading U.S. researcher Professor D.A. Brenner, recently elected editor-in-chief of the journal *Gastroenterology*.

The reports given at the two-day meeting were valuable in furthering our understanding of the complicated signaling system involved in hepatocyte differentiation, proliferation, and apoptosis. Progress in this field is rapid, and another symposium on the same theme will be held in the near future. We believe that these proceedings are useful in summarizing current information on this important topic.

The Organizing Committee would like to express special thanks to all participants and to the Viral Hepatitis Research Foundation of Japan for its continuing financial support.

ORGANIZING COMMITTEE OF THE YAMAGUCHI SYMPOSIUM ON LIVER DISEASE
> Kiwamu Okita, M.D., Yamaguchi University, Ube
> Kenichi Kobayashi, M.D., Kanazawa University, Kanazawa
> Masamichi Kojiro, M.D., Kurume University, Kurume
> Masao Omata, M.D., The University of Tokyo, Tokyo
> Norio Hayashi, M.D., Osaka University, Osaka

GENERAL SECRETARY
> Isao Sakaida, M.D., Yamaguchi University, Ube

Table of Contents

List of Participants

Akita, Kuniharu

First Department of Internal Medicine
Gifu University School of Medicine
Gifu, Japan

Brenner, David A.

University of North Carolina at Chapel Hill
North Carolina, USA

Fukumoto, Yohei

Department of Molecular Science & Applied Medicine
Gastroenterology & Hepatology
Yamaguchi University School of Medicine
Yamaguchi, Japan

Hayashi, Norio

Molecular Therapeutics
Osaka University Graduate School of Medicine
Osaka, Japan

Hino, Keisuke

Department of Molecular Science & Applied Medicine
Gastroenterology & Hepatology
Yamaguchi University School of Medicine
Yamaguchi, Japan

Hoshida, Yujin

Department of Gastroenterology
Graduate School of Medicine, The University of
Tokyo
Tokyo, Japan

Ido, Akio

Department of Internal Medicine II
Miyazaki Medical College
Miyazaki, Japan

Kato, Naoya

Department of Gastroenterology
Graduate School of Medicine, The University of
Tokyo
Tokyo, Japan

Kawata, Sumio	Second Department of Internal Medicine Yamagata University School of Medicine Yamagata, Japan
Kayano, Kozo	Department of Molecular Science & Applied Medicine Gastroenterology & Hepatology Yamaguchi University School of Medicine Yamaguchi, Japan
Kobayashi, Kenichi	First Department of Internal Medicine Faculty of Medicine, Kanazawa University Ishikawa, Japan
Kodama, Takahiro	Department of Internal Medicine Yamaguchi Central Hospital Yamaguchi, Japan
Koga, Hironori	Second Department of Medicine Kurume University School of Medicine Fukuoka, Japan
Kojiro, Masamichi	Department of Pathology Kurume University School of Medicine Fukuoka, Japan
Nakamoto, Yasunari	First Department of Internal Medicine Faculty of Medicine, Kanazawa University Ishikawa, Japan
Nakanishi, Toshio	Department of General Medicine Hiroshima University Hiroshima, Japan
Okita, Kiwamu	Department of Molecular Science & Applied Medicine Gastroenterology & Hepatology Yamaguchi University School of Medicine Yamaguchi, Japan
Okuno, Masataka	First Department of Internal Medicine Gifu University School of Medicine Gifu, Japan
Omata, Masao	Department of Gastroenterology Graduate School of Medicine, The University of Tokyo Tokyo, Japan

Saito, Takafumi

Second Department of Internal Medicine
Yamagata University School of Medicine
Yamagata, Japan

Sakaida, Isao

Department of Molecular Science & Applied Medicine
  Gastroenterology & Hepatology
Yamaguchi University School of Medicine
Yamaguchi, Japan

Sasaki, Yutaka

Molecular Therapeutics
Osaka University Graduate School of Medicine
Osaka, Japan

Sata, Michio

Second Department of Medicine
Kurume University School of Medicine
Fukuoka, Japan

Sugiyama, Toshihiro

Department of Biochemistry
Akita University School of Medicine
Akita, Japan

Tanikawa, Kyuichi

International Institute for Liver Research
Kurume, Japan

Terai, Shuji

Department of Molecular Science & Applied
  Medicine, Bioregulation
Yamaguchi University School of Medicine
Yamaguchi, Japan

Tsubouchi, Hirohito

Department of Internal Medicine II
Miyazaki Medical College
Miyazaki, Japan

HHM: A Dominant Inhibitory Helix-Loop-Helix Protein Associated with Liver-Specific Gene Expression and Liver Stem Cells

Shuji Terai[1,2], Snorri S. Thorgeirsson[3], and Kiwamu Okita[2]

Summary. The helix-loop-helix (HLH) family of transcriptional regulatory proteins are key regulators in numerous organ development. We had identified a novel HLH factor, human homologue of Maid (HHM), from a human fetal liver cDNA library. HHM is composed of 360 amino acids (aa). The HHM has an HLH region (150–184 aa) with a leucine zipper motif (240–260 aa) lacking the basic region. The homology of the HLH region for HHM shared 82% identity with that for maternal Id-likemolecule (Maid), but showed low homology with other dominant inhibitory HLH proteins such as Ids. HHM is expressed at a high level in fetal liver as well as in brain, placenta, bone marrow, and lung and is transiently high during liver stem cell activation in the 2-acetylaminofluorene/partial hepatectomy (AAF/PH) model. HHM also inhibited luciferase gene activation induced by the hepatic nuclear factor 4 (HNF-4) promoter. These results suggest that HHM is a novel type of dominant inhibitory HLH protein that might be associated with liver-specific gene expression and liver stem cells.

Key words. HLH transcription factor, Liver stem cell, Liver-specific gene expression, Liver development

Introduction

The helix-loop-helix (HLH) family of transcriptional regulatory proteins are key regulators of various processes of organ development [1,2]. HLH proteins are classified into seven groups on the basis of tissue distribution, dimerization capability, and DNA binding [2]. Class I HLH proteins, such as E12 and E47, are also called E proteins. These class I HLH proteins are ubiquitously expressed in many tissues and form either homo- or heterodimers [1,2]. The specific DNA-binding site of class I proteins is known as E-box, having the consensus sequence (CANNTG) [3]. Class II HLH proteins, such as MyoD, myogenin, Atonal, and NeuroD/BETA2, show a tissue-specific

[1] Department of Molecular Science & Applied Medicine, Bioregulation, [2] Department of Molecular Science & Applied Medicine, Gastroenterology & Hepatology, Yamaguchi University School of Medicine, 1-1-1 Minami-Kogushi, Ube, Yamaguchi 755-8505, Japan
[3] Laboratory of Experimental Carcinogenesis, National Cancer Institute, National Institutes of Health, Bethesda, MD, USA

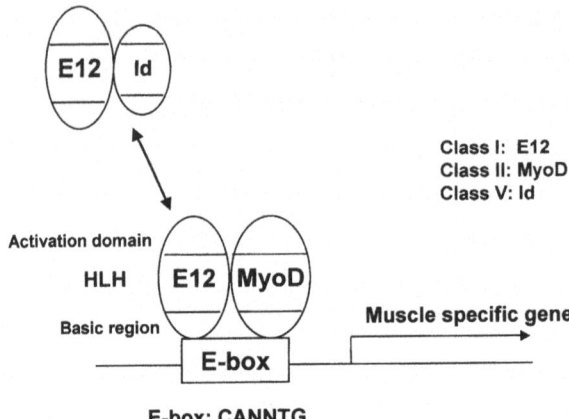

E-box: CANNTG

Fig. 1. Regulation of muscle cell differentiation by E12/MyoD and Id. The *E12/MyoD* heterodimer binds to E-box and regulates muscles cell differentiation. The dominant inhibitory helix-loop-helix (HLH) protein Id binds to E12 and inhibits the binding of the E12/MyoD heterodimer to E-box. As a result, Id acts as an inhibitor of differentiation

expression pattern [2]. The heterodimers of class I and II HLH proteins bind to the E-boxes that are found in a number of cell type-specific promoters and enhancers and regulate different developmental pathways, such as myogenesis (Fig. 1) [1–3]. Class III HLH proteins include the Myc family of transcription factors, TFE-3, SREBP-1, and the microphthalmia-associated transcription factor Mi [2,4,5]. Proteins of this class contain a leucine zipper (LZ) adjacent to the HLH motif [2]. Class IV HLH proteins include Mad, Max, and Mxi, which are capable of dimerizing with the Myc proteins or with one another [6,7].

Class V HLH protein lack the basic domain and consequently do not bind to DNA; these include Id and emc, and are negative regulators of class I and class II HLH proteins [8,9]. Class VI HLH proteins have a specific feature, a proline in their basic region. *Drosophila* proteins Hairy and Enhancer of split are included in this class [10,11]. The class VII HLH proteins, characterized by the presence of the bHLH-PAS domain, include aromatic hydrocarbon receptor (AHR) and the AHR nuclear translocator (Arnt) [12].

The Id family is a dominant inhibitory HLH protein [8,9]. Id proteins, inhibitors of DNA binding and differentiation, are negative regulators of bHLH [2,8,9]. All Id proteins contain the HLH domain and heterodimerize with bHLH factors but lack the basic region; as a result, the heterodimers lose the DNA-binding activity. This disruption of DNA binding is closely associated with the inhibition of differentiation [2,8,9]. The four Id proteins, Id1–4, that have been identified are highly homologus in the HLH regions and show similar affinities for various kinds of E protein [2,9]. An Id-like molecule, Maid, has been previously identified and shown to be a potentially dominant inhibitory HLH protein as exemplified by inhibition of E12/MyoD dimerization, but the precise function has not been evaluated [13]. Similar to the Id proteins, Maid also lacks the basic region, but differs from the Id proteins in harboring an LZ motif.

Little information exists on the potential impact of HLH proteins on early liver development or on activation and differentiation of liver stem cells. In contrast, in pancreatic development a number of pancreas-specific genes have been shown to contain E-box elements that regulate their tissue-specific expression [14–16]. We have found E-box sequences in promoters/enhancers of a number of genes that both regulate and characterize liver-specific gene expression; these include hepatic nuclear factor (HNF)-1α, HNF-3α and 3β, HNF-4, HNF-6, and alpha-fetoprotein (AFP) as well as the intergenic enhancer between albumin and AFP genes [17–21]. From these results, we tried to identify partners for E12 that might act as potential regulators of liver-specific gene expression. A yeast two-hybrid screen using an interaction trap system with the HLH region of E12 protein fused to DNA-binding protein LexA (termed LexA-E12) as bait was performed. As a result, we identified the novel HLH factor, HHM, from a human fetal liver cDNA library [22].

Structure of HHM

The HHM protein is composed of 360 amino acids and has a predicted molecular mass of 40.8 kDa (Fig. 2). The HHM protein contains a region of 35 amino acids (positions 150–184 aa) with a similarity of HLH domain and 22 amino acids of putative LZ motif (240–261 aa) (Fig. 2). Similar to the mouse Maid protein, HHM lacks the basic DNA-binding domain, suggesting that HHM might be a dominant inhibitory HLH protein. In comparison to HHM, the Id family is smaller and lacks the LZ motif. The HLH domains for HHM shared 82% identity with that for Maid but had low homology with those for Ids, suggesting that HHM might be a distant member of the Id family and belong to class V [2,13,22].

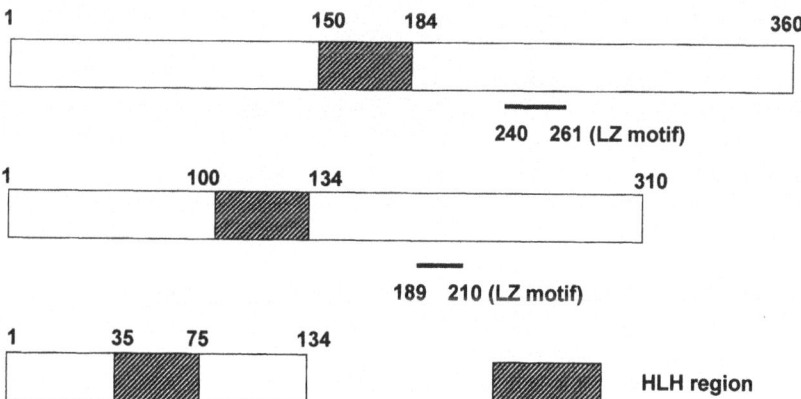

Fig. 2. Structure of the human homologue of Maid (HHM) protein. Schematic illustration of HHM, Maid, and Id2. The HHM protein is composed of 360 amino acids. HHM and Maid proteins are longer than Id2; HHM and Maid proteins have the leucine zipper (*LZ*) motif

Expression Profile of HHM

The level of Maid transcripts in adult mouse tissues was reported to be high in ovary, intestine, muscle, and brain, whereas low hybridization was observed in testis, heart, kidney, and spleen. Also, a low level of hybridization was observed with lung and liver RNA after long exposure of the blot [13]. We had analyzed the expression profile of HHM using the Multiple Tissue Expression Array (obtained from Clontech, CA, USA). Abundant expression of HHM mRNA was observed in bone marrow, brain, pituitary gland, adrenal gland, thyroid gland, spleen, lung, heart, and placenta. Interestingly, abundant HHM mRNA transcripts were also observed in fetal but not in adult liver [22].

Expression of HHM mRNA During Liver Regeneration

It was of interest that abundant HHM mRNA expression was specifically observed in fetal but not in adult liver [22]. This observation suggested that HHM might be involved in regulating both growth and differentiation in the liver. Our initial approach to addressing this issue was to analyze the HHM gene expression during liver regeneration by utilizing two different rat models, the classical partial hepatectomy (PH) model in which 70% of the liver mass is surgically removed and the restoration of the liver takes place from the remaining hepatocytes, and the combination of PH and treatment with 2-acetylaminofluorene (the AAF/PH model) representing stem cell-supported liver regeneration [23]. In the PH model, the peak level of HHM mRNA was observed during maximum DNA synthesis at 24–36 h after the operation, whereas the highest HHM mRNA expression was observed between 13 and 16 days in the AAF/PH model (Fig. 3). In the AAF/PH model, the time period of maximum HHM expression coincided with the period at which foci of newly generated basophilic hepatocytes were abundant [22]. Furthermore, in situ hybridization analysis revealed that HHM mRNA was preferentially expressed in the basophilic foci in the AAF/PH model [22].

HHM Inhibits Transcriptional Activity Induced by the HNF-4 Promoter

We analyzed the inhibitory function of HHM in HepG2 cells by introduction of the plasmid pGL3-HNF4 E-box carrying the luciferase reporter gene fused to HNF-4 promoter and effector plasmids expressing HHM, ΔHHM (1–149), Id2, and pcDNA-FLAG as a control. The 5'-region of HNF-4 promoter from −363 to +48 bp is sufficient to drive liver-specific expression of the luciferase gene in HepG2 cells [24]. We found that two E-boxes exist in these regions (from −179 to −174 bp, CAGTTG; from −161 to −156 bp, CAGGTG). The introduction of HHM efficiently suppressed HNF-4 promoter-induced luciferase gene activation. However, ΔHHM (1–149) and Id2 did not suppress luciferase gene activation [22].

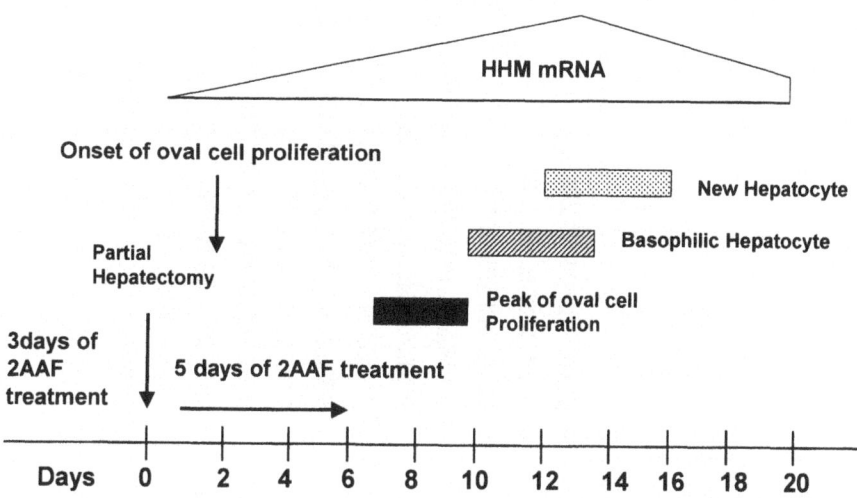

Fig. 3. HHM mRNA expression in the 2-acetylaminofluorene/partial hepatectomy (AAF/PH) model. In this model can be seen the generation of the oval cell and sequential differentiation into the basophilic hepatocyte and hepatocyte. HHM mRNA is elevated from the period when the proliferation of the oval cell starts. The peak value *HHM mRNA* was seen at the time of the appearance of the basophilic hepatocytes. The HHM mRNA transcript was specifically seen in the foci of basophilic hepatocytes by in situ hybridization

Discussion

We have identified a novel inhibitory HLH factor that interacts with E12, displays extensive homology with Maid, and has consequently been named human homologue of Maid [22]. Maid was previously shown to be an inhibitory HLH protein that impeded the dimerization of E12/MyoD, but the precise function of Maid had not been identified [13]. Physical interaction of HHM with bHLH protein, E12, was demonstrated by coprecipitation assay and a glutathione S-transferase-pulldown experiment [22]. Both HHM and Maid contain an LZ region in addition to the HLH motif, and this feature provides a distinct structural difference from the Id proteins (Figure 2). However, based on a recent review by Massari and Murre [2], HHM and Maid may be included with the Id family in class V. HHM is able to heterodimerize with E12 through the HLH region, but these heterodimeric complexes are functionally inactive. By titrating out E12 and MyoD, HHM, similar to Id2, inhibits DNA binding of the E12-MyoD heterodimer; however, inhibition by HHM was less than that by Id2. It was noteworthy that the HLH domain of HHM is quite different from that of the Id family [22].

Although little is known about the role of HLH proteins in the regulation of liver-specific gene expression, a number of hepatic transcription factors and liver-specific genes including HNF-1α, HNF-3α and -β, HNF-4, HNF-6, and AFP contain E-boxes in their promoters and enhancers [17–21]. It is, however, known that a number of genes

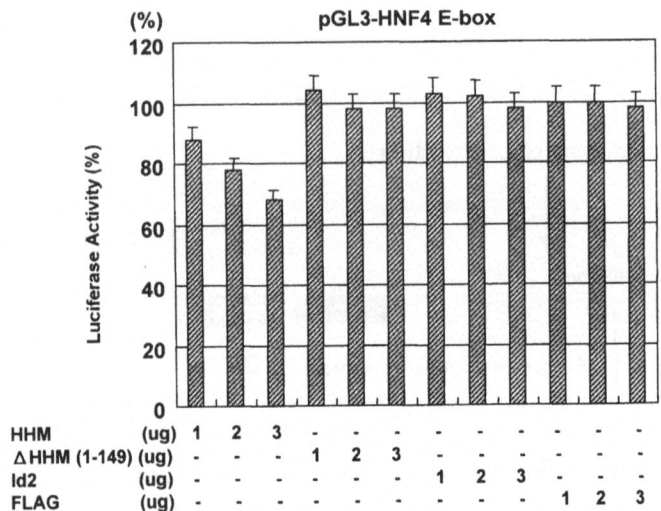

Fig. 4. HHM inhibits the transcriptional activation of hepatic nuclear factor (*HNF-4*). Luciferase assay. HepG2 cells were transiently cotransfected with 1.5 μg pGL3-HNF-4 E-box, 75 ng of pRL-TK, together with indicated expression vector encoding *HHM*, *ΔHHM(1–149)*, *Id2*, and control pcDNA-*FLAG*. The combination of transfected plasmids (in μg) is shown in the *lower portion*. HHM efficiently suppressed HNF-4 promoter-induced luciferase gene activation. Neither ΔHHM(1–149) nor Id2 suppressed luciferase gene activation

selectively expressed in the pancreas require E-box sites for proper expression [15,16]. The α and β cells express both HEB and E2A and heterodimerize to a class II protein BETA2/NeuroD [15]. Targeted deletion of BETA2/NeuroD in mice results in severe diabetes and death at 3–5 days of age, indicating that the BETA2/NeuroD gene is critical for the normal development of several specialized cell types arising from gut endoderm [16]. A number of the HNFs that harbor E-boxes are sequentially expressed during liver development and following activation of liver stem cells in adult liver (Fig. 4). The early hepatoblasts derived from the gut endoderm are bipotential progenitors capable of differentiating into both bile epithelial cells and hepatocytes. The hepatoblasts undergo distinct phases of differentiation as they progress along the hepatocytic lineage, and each transition is characterized by the appearance of regulatory transcription factors [25]. Similarly, during activation and differentiation of the stem cells in adult rat liver via the hepatocyte lineage, a set of transcription factors are sequentially expressed (Fig. 5) [26,27]. For example, at the early stage of stem cell activation both the small and large bile ducts start to express HNF-1α and -β, HNF-3α, and C/EBP-α and -β, but not HNF-4. At the later stages these transcription factors are also highly expressed in the proliferating oval cells, the early progenitors from the stem cell compartment. However, HNF-4 is first observed when the oval cells differentiate morphologically and functionally into hepatocytes and form basophilic foci [25,26]. The observation that HHM is transiently expressed during stem cell-driven

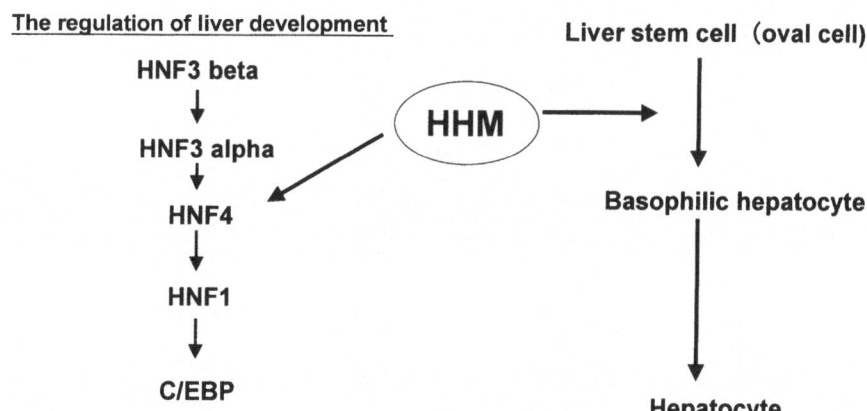

Fig. 5. Function of HHM. Schematic illustration of the function of HHM. HNFs and C/EBP regulate liver development. Especially, HHM might regulate HNF-4. HHM might have an important role at the time when liver stem cells (oval cells) differentiate into basophilic hepatocytes

regeneration of the liver at the stage at which the early basophilic foci of hepatocytes start to appear is of potential interest. HHM might control the regulation of this stage. We also analyzed how HHM regulates the luciferase gene activation induced by the HNF-4 promoter [24]. We used the HNF-4 promoter, which is sufficient to induce liver-specific gene expression in HepG2 cells [24]. Two E-boxes are present in this promoter region of HNF-4. Interestingly, HHM suppressed luciferase gene activation induced by HNF-4 promoter (see Fig. 4). However, Id2 and ΔHHM (1–149) could not suppress the luciferase gene activation induced by HNF-4. We cannot address the molecular mechanism showing why HHM but not Id2 only suppressed luciferase activity, but this result demonstrated that HHM is involved in the regulation of HNF-4. That the HHM transcripts are specifically located over the basophilic foci and that HHM regulates transcriptional activity of HNF-4 demonstrated the important role of HHM in liver development (see Fig. 5).

On the other hand, two other groups independently identified the same gene with HHM; they called the genes, *DIP1* and *GCIP* [28,29]. *DIP1* was identified by the yeast two-hybrid screen using cyclin D_1 as a bait protein. GCIP was also identified as an interacting protein with Grap2, which is a novel adaptor protein identified as a Gab-1 docking protein by the yeast two-hybrid screen. The precise function of this protein, HHM/GCIP/DIP1, has not yet been identified. As shown in Fig. 6, HHM might act as a transcriptional regulatory protein that might interact with cyclin D_1 or an unknown liver-specific bHLH transcription factor. Now, we are analyzing the in vivo function of HHM in the regulatory mechanism for proliferation and differentiation by analyzing interacting protein with HHM.

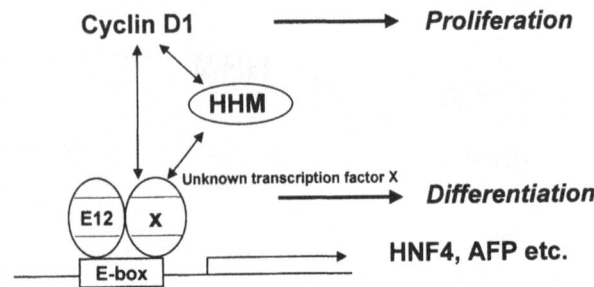

Cell Cycle Regulation

Cyclin D1 ⟶ *Proliferation*

HHM

E12 X Unknown transcription factor X ⟶ *Differentiation*

E-box ⟶ **HNF4, AFP etc.**

Growth Factor

HGF, SCF, TGF-beta, EGF

Fig. 6. HHM regulates cell proliferation and differentiation by binding with various proteins. HHM might interact with cyclin D_1. The HHM–cyclin D1 heterodimer might regulate cell proliferation. On the other hand, HHM might also interact with an unknown transcription factor X and regulate a liver-specific gene such as HNF-4 or alpha-fetoprotein. HHM could act as the regulator of differentiation. Several kinds of growth factors such as hepatocyte growth factor, stem cell factor, transforming growth factor-beta, and epidermal growth factor might modulate the function of HHM

References

1. Olson EN, Klein WH (1994) bHLH factors in muscle development: deadlines and commitments, what to leave in and what to leave out. Genes Dev 8:1–8
2. Massari ME, Murre C (2000) Helix-loop-helix proteins: regulators of transcription in eucaryotic organisms. Mol Cell Biol 20:429–440
3. Murre C, McCaw PS, Vaessin H, Caudy M, Jan LY, Jan YN, et al (1989) Interactions between heterologous helix-loop-helix proteins generate complexes that bind specifically to a common DNA sequence. Cell 58:537–544
4. Henthom PS, Stewart CC, Kadesch T, Puck JM (1991) The gene encoding human TFE3, a transcriprion factor that binds the immunoglobulin heavy-chain enhancer, maps to Xp11.22. Genomics 11:374–378
5. Zhao GQ, Zhao Q, Zhou X, Mattei MG, de Crombrugghe B (1993) TFEC, a basic helix-loop-helix protein, forms heterodimers with TFE3 and inhibits TFE3-dependent transcription activation. Mol Cell Biol 13:4505–4512
6. Ayer DE, Kretzner L, Eisenman RN (1993) Mad: a heterodimeric partner for Max that antagonizes Myc transcriptional activity. Cell 72:211–222
7. Blackwood EM, Eisemnan RN (1991) Max: a helix-loop-helix zipper protein that forms a sequence-specific DNA-binding complex with Myc. Science 251:1211–1217
8. Ellis HM, Spann DR, Posakony JW (1990) Extramacrochaetae, a negative regulator of sensory organ development in *Drosophila*, defines a new class of helix-loop-helix proteins. Cell 61:27–38
9. Norton JD, Deed RW, Craggs G, Sablitzky F (1998) Id helix-loop-helix proteins in cell growth and differentiation. Trends Cell Biol 8:58–65

10. Klambt C, Knust E, Tietze K, Campos-Ortega JA (1989) Closely related transcripts encoded by the neurogenic gene complex enhancer of split of Drosophila melanogaster. EMBO J 8:203–210
11. Rushlow CA, Hogan A, Pinchin SM, Howe KM, Lardelli M, Ish-Horowicz D (1989) The Drosophila hairy protein acts in both segmentation and bristle patterning and shows homology to N-myc. EMBO J 8:3095–3103
12. Crews ST (1998) Control of cell lineage-specific development and transcription by bHLH-PAS proteins. Genes Dev 12:607–620
13. Hwang SY, Oh B, Fuchtbauer A, Fuchtbauer EM, Johnson KR, Solter D, et al (1997) Maid: a maternally transcribed novel gene encoding a potential negative regulator of bHLH proteins in the mouse egg and zygote. Dev Dyn 209:217–226
14. German MS, Blanar MA, Nelson C, Moss LG, Rutter WJ (1991) Two related helix-loop-helix proteins participate in separate cell-specific complexes that bind the insulin enbancer. Mol Endocrinol 5:292–299
15. Shieh SY, Tsai MJ (1991) Cell-specific and ubiquitous factors are responsible for the enhancer activity of the rat insulin II gene. J Biol Chem 266:16708–16714
16. Naya FJ, Huang HP, Qiu Y, Mutoh H, DeMayo FJ, Leiter AB, et al (1997) Diabetes, defective pancreatic morphogenesis, and abnormal enteroendocrine differentiation in BETA2/neuroD-deficient mice. Genes Dev 11:2323–2334
17. Sakai M, Morinaga T, Urano Y, Watanabe K, Wegmann TG, Tamaoki T (1985) The human alpha-fetoprotein gene. Sequence organization and the 5′ flanking region. J Biol Chem 260:5055–5060
18. Nakabayashi H, Hashimoto T, Miyao Y, Tjong KK, Chan J, Tamaoki T (1991) A position-dependent silencer plays a major role in repressing alpha-feto-protein expression in human hepatoma. Mol Cell Biol 11:5885–5893
19. Bernier D, Thomassin H, Allard D, Guertin M, Hamel D, Blaquiere M, et al (1993) Functional analysis of developmentally regulated chromatin-hypersensitive domains carrying the alpha 1-fetoprotein gene promoter and the albumin/alpha 1-fetoprotein intergenic enhancer. Mol Cell Biol 13:1619–1633
20. Rastegar M, Szpirer C, Rousseau GG, Lemaigre FP (1998) Hepatocyte nuclear factor 6: organization and chromosomal assignment of the rat gene and characterization of its promoter. Biochem J 334:565–569
21. Taraviras S, Monaghan AP, Schutz G, Kelsey G (1994) Characterization of the mouse HNF-4 gene and its expression during mouse embryogenesis. Mech Dev 48:67–79
22. Terai S, Aoki H, Ashida K, Thorgeirsson SS (2000) Human homologue of Maid: a dominant inhibitory helix-loop-helix protein associated with liver specific gene expression. Hepatology 32:357–366
23. Golding M, Sarraf CE, Lalani EN, Anilkumar TV, Edwards RJ, Nagy P, et al (1995) Oval cell differentiation into hepatocytes in the acetylaminofluorene-treated regenerating rat liver. Hepatology 22:1243–1253
24. Zhong W, Mirkovitch J, Darnell JE Jr (1994) Tissue-specific regulation of mouse hepatocyte nuclear factor 4 expression. Mol Cell Biol 14:7276–7284
25. Grisham JW, Thorgeirsson SS (1997) Liver stem cells. In: Potten CS (ed) Stem cells. Academic Press, London, pp 233–282
26. Zaret K (1998) Early liver differentiation: genetic potentiation and multilevel growth control. Curr Opin Genet Dev 8:526–531
27. Nagy P, Bisgaard HC, Thorgeirsson SS (1994) Expression of hepatic transcription factors during liver development and oval cell differentiation. J Cell Biol 126:223–233
28. Yao Y, Doki Y, Jiang W, Imoto M, Venkatraj VS, Warburton D, Santella RM, Lu B, Yan L, Sun XH, Su T, Luo J, Weinstein IB (2000) Cloning and characterization of DIP1, a novel protein that is related to the Id family of proteins. Exp Cell Res 257(1):22–32
29. Xia C, Bao Z, Tabassam F, Ma W, Qiu M, Hua S, Liu M (2000) GCIP, a novel human grap2 and cyclin D interacting protein, regulates E2F-mediated transcriptional activity. J Biol Chem 275(27):20942–20948

Expression of Musashi-1 Antigen, A Neural Stem Cell RNA-Binding Protein, in the Liver Tissues of Patients with Hepatitis

TAKAFUMI SAITO[1], HIROKO MITSUI[2], HISAYOSHI WATANABE[1], YUKI TERUI[1],
TADASHI TAKEDA[1], MASANORI AOKI[1], KOJI SAITO[1], RYUICHI NAGASHIMA[1],
HIROAKI TAKEDA[1], HIROYUKI MISAWA[1], HITOSHI TOGASHI[1],
MITSUNORI YAMAKAWA[2], and SUMIO KAWATA[1]

Summary. Musashi-1 is a gene encoding the RNA-binding protein involved in regulating asymmetrical division in neural stem cells. As Musashi-1 antigen (Msi-1) can be a potent marker for liver stem cells, we used it as an immunological tool to investigate the mechanisms involved in liver reconstruction after injury. The expression of CD 68 antigen (CD 68), a marker for macrophages, was also analyzed. Immunohistochemical studies were performed on liver biopsy specimens from patients with various liver diseases, including 9 with acute hepatitis, 13 with chronic hepatitis, and 5 with liver cirrhosis. Msi-1 was expressed predominantly in the cytoplasm of CD 68-positive macrophages, which have the potential for activation or are already activated. These cells were found in necrotic or inflamed parenchyma or in the portal area, especially in the acute hepatitis specimens. Most of the Kupffer cells (resident liver macrophages) were located in noninflamed areas and were positive for CD 68 but negative for Msi-1. We also observed Msi-1 expression on oval-shaped cells in the inflamed portal area, although further analysis is necessary to determine whether these cells were liver stem cells. The asymmetrical division that occurs in such stem cells and is regulated by Msi-1 may play a role in promoting recovery from hepatitis and subsequent reconstruction of the liver by activating the mononuclear phagocyte system, as well as by replacing damaged liver cells with new ones derived from stem cells.

Key words. Musashi-1, Stem cell, Oval cell, Macrophage, Hepatitis

Introduction

Many types of liver injury induce biological responses that promote the recovery and subsequent reconstruction of the injured liver. Hepatocytes can proliferate independently by division after the loss of liver cells, as is often observed after partial hepatectomy [1,2]. An alternative mechanism in which liver stem cells that subsequently differentiate into hepatocytes and cholangiocytes are produced is involved in recon-

[1] Second Department of Internal Medicine, [2] First Department of Pathology, Yamagata University School of Medicine, 2-2-2 Iidanishi, Yamagata 990-9585, Japan

11

struction after liver damage [3,4]. The most likely candidates for liver stem cells are oval cells, which possess both hepatocyte and biliary cell markers [5–8]. However, to date, there have been no reports of oval cells emerging in the liver tissue of humans with hepatitis. The mononuclear phagocyte system (MPS) also plays an important role in repairing the injured liver after inflammation because its dynamic function helps to regulate the proper reconstruction of the liver [9]. However, the characteristics of the cells associated with the MPS (for example, their origins and modes of differentiation) are not fully understood. Thus, new approaches to the investigation of the mechanisms involved in liver reconstruction, with particular reference to stem cells, are required.

Musashi-1 antigen (Msi-1) is a neural RNA-binding protein that is specifically expressed in the precursor cells of the developing mouse central nervous system (CNS) [10]. It is closely involved in the regulation of asymmetrical cell division, which generates differentiated cells. Msi-1 can be a potent marker for both liver stem cells, which differentiate into new hepatocytes, and liver-infiltrating stem cells, which are associated with the MPS. We therefore carried out an immunohistochemical study to investigate the expression of Msi-1 in the liver tissue of patients with various liver diseases. The expression of CD 68 antigen (CD 68), a well-established marker for macrophages, was also studied to distinguish between the different types of Msi-1-positive cells. Our results could indicate a future direction for research into the involvement of liver stem cells in liver reconstruction.

Materials and Methods

Liver Tissues

Liver biopsy specimens from 27 patients with liver disease were examined. The liver tissue was obtained by diagnostic needle biopsy, after obtaining informed consent. The histological features of the specimens were classified according to international standard criteria [11], and included acute hepatitis (AH) in 9 patients, chronic hepatitis (CH) in 13, and liver cirrhosis (LC) in 5.

Antibodies

A rat monoclonal antibody against Msi-1 (anti-Msi-1) was kindly provided by Dr. Hideyuki Okano (Osaka University Medical School, Osaka, Japan). A mouse monoclonal antibody against CD 68 (anti-CD 68) was purchased from Dako (Kyoto, Japan).

Immunohistochemistry

The liver biopsy specimens were fixed with 10% formalin and embedded in paraffin. Four-micrometer-thick serial sections were prepared, deparaffinized, and washed extensively with phosphate-buffered saline (PBS), pH 7.2, before immunostaining. The Msi-1 antigen was unmasked by microwaving at 800 W for 20 min in 0.01 M citrate buffer, pH 6.0. The CD 68 antigen was unmasked by the treatment of 0.1% trypsin in

PBS for 30 min. An avidin–biotin complex immunoperoxidase staining technique was employed (Vectastain ABC kit; Vector, Burlingame, CA). After inhibiting the endogenous peroxidase activity of the liver tissue with 0.3% hydrogen peroxide solution for 30 min at 4°C, the sections were incubated with anti-Msi-1 (1:200) and anti-CD68 (1:50) in a moist chamber overnight at 4°C, followed by incubation with the biotinylated secondary antibodies and the avidin–biotin complex according to the manufacturer's instructions. The reaction products were stained with a 0.02% solution of 3,3′-diaminobenzidine in 0.05 M Tris-HCl buffer, pH 7.2, containing 0.01% hydrogen peroxide. The slides were counterstained with methyl green for Msi-1 and hematoxylin for CD 68, dehydrated, and mounted.

Statistical Analysis

Differences in the numbers of Msi-1-positive cells between the various types of liver disease were determined using Student's t test for unpaired samples. Differences with P values <0.05 were considered significant.

Results

Numbers of Msi-1-Positive Cells in Relation to Liver Histology

Although Msi-1 antigen was detected in all types of liver disease, it was particularly strongly expressed in the AH specimens. The mean numbers of positive cells in specimens with AH, CH, and LC were 118.1 ± 90.9, 4.7 ± 4.8, and 6.2 ± 7.3 (mean ± SD), respectively. The number of positive cells was significantly higher in the specimens with AH than in those with CH ($P < 0.01$) or LC ($P < 0.05$).

Expression and Localization of Msi-1 and CD 68 in the Liver Tissue of Patients with Hepatitis

Msi-1-positive cells were detected mainly in the inflamed area of the liver. Particularly large numbers of these cells were found in areas of the liver parenchyma with focal or confluent necrosis and in the swollen portal areas in the AH specimens. Msi-1 was mainly expressed on cells associated with the MPS. The morphology of these cells varied according to their origins. Inflammation results in activation of the MPS, and macrophages resulting from the division of monocytes derived from progenitor cells within the bone marrow are supplied to the inflamed area.

Other macrophages may be produced by extramedullary hematopoiesis, and these are often observed in the inflamed liver. Msi-1 was detected in the cytoplasm of such macrophages, which either have the potential for activation or are already activated (Fig. 1a). These Msi-1-positive cells were swollen by phagocytosis and could also be detected as cells with ceroid granules in hematoxylin and eosin (HE)-stained sections. We confirmed that these cells were macrophages by positive staining for CD 68 (Fig. 1b). However, the localization of these Msi-1-positive macrophages was not consistent with the normal distribution of CD 68-positive cells in the liver. Most of the Kupffer cells (resident liver macrophages) were negative for Msi-1 but positive for CD 68 and

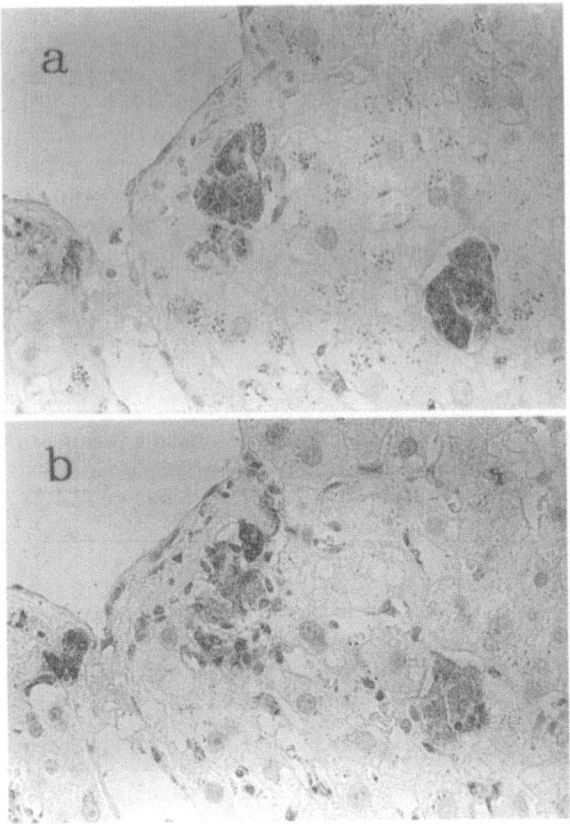

Fig. 1. Musashi-1 antigen (Msi-1) was detected in the cytoplasm of activated macrophages, which appeared swollen due to phagocytosis (**a**). These cells were confirmed to be macrophages by positive staining for CD 68 antigen (CD 68) in serial sections (**b**). ×400

were found in noninflamed areas (Fig. 2). We also observed Msi-1 expression on oval-shaped cells in the inflamed portal area (Fig. 3), although further analysis is necessary to determine whether these cells are liver stem cells.

Discussion

Msi-1 expression is observed not only in neural precursor cells, which are capable of generating both neurons and glia during the embryonic development of the central nervous system [10], but also in early hepatocytes during liver development in the mouse (unpublished data). Msi-1 can also be detected in postnatal neural precursor cells, which are closely involved in the asymmetrical divisions that generate differentiated cells [12]. Thus, even in adults, this antigen can act as a marker for the liver stem cells that are induced to reconstruct the liver architecture during an ongoing

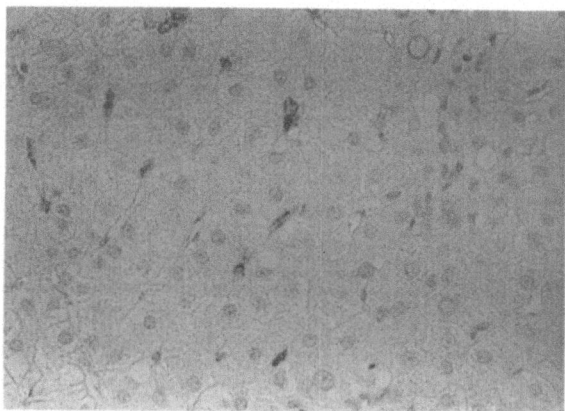

Fig. 2. Most of the Kupffer cells (the resident liver macrophages) were located in noninflamed areas and were positive for CD 68. Note that these cells were negative for Msi-1. ×400

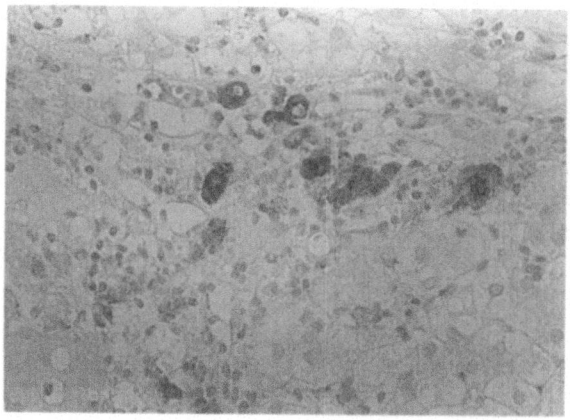

Fig. 3. Msi-1 was expressed on oval-shaped cells in the inflamed portal area. ×400

inflammatory process. In this study, we showed for the first time that Msi-1 is expressed in the liver tissue of adult humans with hepatitis.

Liver stem cells can be divided into two types: hepatocytes and cholangiocytes. Oval cells share some of the characteristics of liver stem cells, because in vivo study has shown that they can be transformed into albumin-producing hepatocytes in the rat liver [13]. In HE sections of human liver tissue with hepatitis, a group of small hepatocytes with a high nuclear-to-cytoplasmic ratio, whose morphology closely resembles that of oval cells, are found around the inflamed portal area. These hepatocytes are reported to be immature cells because electron microscopic observation revealed only scanty cell organelles [14]. In the present study, we found similar cells that were positive for Msi-1 around the inflamed portal area, although further careful observations are required to allow these cells to be distinguished from the macrophages which are predominantly expressed in this area. Further immunohistochemical studies, including investigations of the stem cell-related antigens c-*kit* [15] and CD 34 [16] by double immunostaining and immununoelectron microscopy, are now in progress to determine whether liver stem cells appear in the liver of adult humans with hepatitis.

During the inflammatory process, the MPS plays an important role in the reconstruction of the liver [9]. It is thought to be activated via several pathways. Once liver inflammation has begun to occur, immature macrophages are found in the liver [17]. These cells are derived from monocytes transformed from granulocyte/macrophage progenitor cells, which in turn originate from stem cells in the bone marrow [18,19]. These immature macrophages have the potential to divide into cells with various functions; some become activated macrophages that work as phagocytes, while others become resident macrophages which eventually settle in the liver and become Kupffer cells [20,21]. Activated macrophages may also be supplied by extramedullary hematopoiesis, which is induced during liver regeneration.

During the present study, Msi-1 was detected in activated macrophages that were assumed to have differentiated from immature macrophages with stem macrophage-like features. Although no Msi-1 was detected in most of the Kupffer cells (which are inactive macrophages not associated with inflammation), detailed observations showed positive Msi-1 staining in some of them (data not shown). As a previous study has shown that Kupffer cells can divide [22], Msi-1 might regulate their differentiation by acting as an "on/off switch" for their activation. These findings suggest that the differentiation of macrophages is associated with activation of the MPS.

Based on our present findings, we have formulated a hypothesis that might explain the development of Msi-1 expression during inflammatory processes involving the liver (Fig. 4). In conclusion, asymmetrical division of stem cells, regulated by the Msi-1 found in these cells, may play a role in promoting recovery from hepatitis and

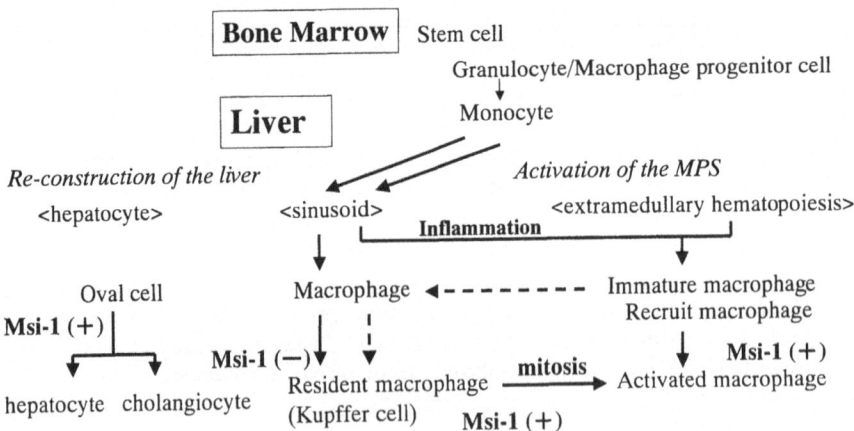

Fig. 4. Hypothesis explaining the development of Msi-1 expression during inflammatory processes involving the liver. *MPS*, mononuclear phagocyte system

subsequent reconstruction of the liver by activating the MPS as well as by replacing damaged liver cells with new ones derived from stem cells.

References

1. Bucher NL (1967) Experimental aspects of hepatic regeneration. N Engl J Med 277: 686–696
2. Rubin EM, Martin AA, Thung SN, et al (1995) Morphometric and immunohistochemical characterization of human liver regeneration. Am J Pathol 147:397–404
3. Thorgeirsson SS (1996) Hepatic stem cells in liver regeneration. FASEB J 10:1249–1256
4. Alison MR, Golding M, Sarraf CE, et al (1996) Liver damage in the rat induces hepatocyte stem cells from biliary epithelial cells. Gastroenterology 110:1182–1190
5. Fausto N (1990) Hepatocyte differentiation and progenitor cells. Curr Opin Cell Biol 2:1036–1042
6. Pack R, Heck R, Dienes HP, et al (1993) Isolation, biochemical characterization, long-term culture, and phenotype modulation of oval cells from carcinogen-fed rats. Exp Cell Res 204:198–209
7. Alison M, Golding M, Lalani EN, et al (1997) Wholesale hepatocytic differentiation in the rat from ductular oval cells, the progeny of biliary stem cells. J Hepatol 26: 343–352
8. Lemire JM, Shiojiri N, Fausto N (1991) Oval cell proliferation and the origin of small hepatocytes in liver injury induced by D-galactosamine. Am J Pathol 139:535–552
9. Arii S, Imamura M (2000) Physiological role of sinusoidal endothelial cells and Kupffer cells and their implication in the pathogenesis of liver injury. J Hepatobiliary Pancreat Surg 7:40–48
10. Sakakibara S, Imai T, Hamaguchi K, et al (1996) Mouse-Musashi-1, a neural RNA-binding protein highly enriched in the mammalian CNS stem cell. Dev Biol 176: 230–242
11. Bianchi L, Groote JD, Desmet VJ, et al (1977) Acute and chronic hepatitis revisited. Review by an international group. Lancet ii:914–919
12. Sakakibara S, Okano H (1997) Expression of neural RNA-binding proteins in the postnatal CNS: implications of their roles in neuronal and glial cell development. J Neurosci 17:8300–8312
13. Yasui O, Miura N, Terada K, et al (1997) Isolation of oval cells from Long-Evans cinnamon rats and their transformation into hepatocytes in vivo in the rat liver. Hepatology 25:329–334
14. Seki S, Sakaguchi H, Kawakita N, et al (1990) Identification and fine structure of proliferating hepatocytes in malignant and nonmalignant liver diseases by use of a monoclonal antibody against DNA polymerase alpha. Hum Pathol 21:1020–1030
15. Fujio K, Hu Z, Evarts RP, et al (1996) Coexpression of stem cell factor and c-kit in embryonic and adult liver. Exp Cell Res 224:243–250
16. Omori N, Omori M, Evarts RP, et al (1997) Partial cloning of rat CD34 cDNA and expression during stem cell-dependent liver regeneration in the adult rat. Hepatology 26:720–727
17. Deimann W, Fahimi HD (1979) The appearance of transition forms between monocytes and Kupffer cells in the liver of rats treated with glucan. J Exp Med 149: 883–897
18. Jordan CT, Lemischka IR (1990) Clonal and systemic analysis of long-term hematopoiesis in the mouse. Genes Dev 4:220–232
19. Wu AM, Till JE, Siminovitch L, et al (1968) Cytological evidence for a relationship between normal hematopoietic colony-forming cells and cells of the lymphoid system. J Exp Med 127:455–462

20. Ukai K, Terashima K, Imai Y, et al (1990) Proliferation kinetics of rat Kupffer cells after partial hepatectomy. Acta Pathol Jpn 40:623–634
21. van Bossuyt H, Wisse E (1988) Structural changes in the rat liver by injection of lipopolysaccharide. Cell Tissue Res 251:205–214
22. Crofton RW, Diesselhoff-den Dulk MMC, van Furth R (1978) The origin, kinetics and characteristics of the Kupffer cells in the normal steady state. J Exp Med 148:1–17

Reconstitution of Hepatic Tissues Using Liver Stem Cells

Toshihiro Sugiyama and Kunihiko Terada

Summary. There is an increasing need for liver transplantation, and the demand has not been sufficiently met. The development of therapies such as liver cell transplantation and an artificial liver to replace liver transplantation is urgently needed. Some clinical applications have been attempted with unsatisfactory results. In the future, it will be necessary to establish more effective therapy by incorporating tissue-engineering techniques to construct hepatic tissue that is capable of high functioning. This article discusses the present state of and prospects for the construction of hepatic tissue utilizing tissue engineering.

Key words. Liver stem cell, Oval cell, Cell transplantation, Liver regeneration, Hepatocyte

Introduction

In recent years, the number of patients with severe hepatopathies who cannot be treated by any therapeutic tactic other than liver transplantation has increased. However, it is difficult to secure sufficient donors for the increased demand for liver transplantation. Therefore, liver cell transplantation or treatment modalities using artificial livers have been attempted in patients with acute hepatic failure or metabolic hepatopathies as novel therapeutic tactics substituting for liver transplantation [1–4]. In these procedures, isolated hepatocyte dispersion is transplanted to the liver parenchyma or filled in as an apparatus to compensate for the metabolic functions of the liver. However, normal liver function cannot yet be sufficiently compensated for by these procedures. In addition to these procedures, another treatment modality in which tissue with structures similar to those of the normal liver is artificially constructed by tissue engineering inside or outside the body to recover higher-level liver functions is currently being developed.

Extracellular matrices are generally required as frameworks to reconstruct or regenerate tissues, in addition to cells constituting the tissues. During skin reconstruction, cultured skin sheets are prepared by disseminating epidermal cells over fibroblasts previously cultured using collagen as the framework. Bone tissues are

Department of Biochemistry, Akita University School of Medicine, 1-1-1 Hondo, Akita 010-8543, Japan reconstructed by culturing bone marrow-derived mesenchymal cells with porous hydroxyapatite with a trabecular structure. In addition, the cartilage is regenerated by adhering chondrocytes to the framework of biodegradable polyglycolic acid mesh or collagen sponge. Because the structures of skin, bone, and cartilage are relatively simple, these procedures are at the point of being put into practical use. In particular, cultured skin sheets have already come onto the market. However, the liver consists of various cell components such as hepatocytes, fat-storing cells of Ito, epithelial cells, and Kupffer cells, as well as consisting of the extracellular matrices such as collagen that support these cell components. Furthermore, the liver consists of regularly arranged lobular structures, including finely complicated courses of blood vessels and bile ducts. Therefore, it has been considered for a long time that the artificial reconstruction of the hepatic tissue is very difficult. In recent years, however, the possibility of the artificial reconstruction of hepatic tissues has been suggested. This chapter introduces some procedures for artificially reconstructing hepatic tissues, and also discusses how to reconstruct hepatic tissues that can be used in the treatment of severe hepatopathies, from the perspective of tissue engineering.

Reconstruction of Hepatic Tissues Using the Primary Culture of Isolated Hepatocytes

After perfusion of collagenase solution through the liver, hepatocytes can be recovered in the precipitate when the cell dispersion is centrifuged at a low speed ($50\,g$ for 1 minute). However, isolated hepatocytes cannot be maintained for a long time when only a single layer of hepatocytes is primarily cultured. Therefore, many researchers have attempted to maintain the functions of hepatocytes as long as possible by adding various cytokines to the culture medium or by culturing hepatocytes with extracellular matrices such as collagen.

Michalopoulos et al. [5] used collagen-coated beads as the framework to reconstruct hepatic tissues from the primary culture of isolated hepatocytes. Briefly, when they repeatedly cultured isolated rat hepatocytes with type I collagen-coated beads in culture medium supplemented with hepatocyte growth factors (HGF) and epidermal growth factors (EGF), cell groups consisting of hepatocytes adhering to the beads were formed. Furthermore, when they embedded these cell groups into Matrigels (cell culture bases containing sarcoma-derived extracellular matrices), hepatocytes proliferated three dimensionally, suggesting the formation of tubular hepatic tissue-like structures (Fig. 1A). It was considered that the hepatic tissue-like structures reconstructed by this culturing procedure maintained some liver function because the expression of albumin, α_1-antitrypsin, and P450 was confirmed in these structures.

Moreover, ectopic in vivo reconstruction of hepatic tissues has been attempted using primary cultured mouse hepatocytes. If tissues structurally similar to those of the liver can be reconstructed in the body by transplanting hepatocytes, exhibition of higher-level liver functions can be expected. Ajioka et al. [6] transferred the vascular epidermal growth factor (VEGF) gene, which encodes vascular growth factors, to hepatocytes beforehand to induce a vascular network that can supply sufficient nutrients and oxygen to reconstructed tissues. As a result, hepatocytes were transformed

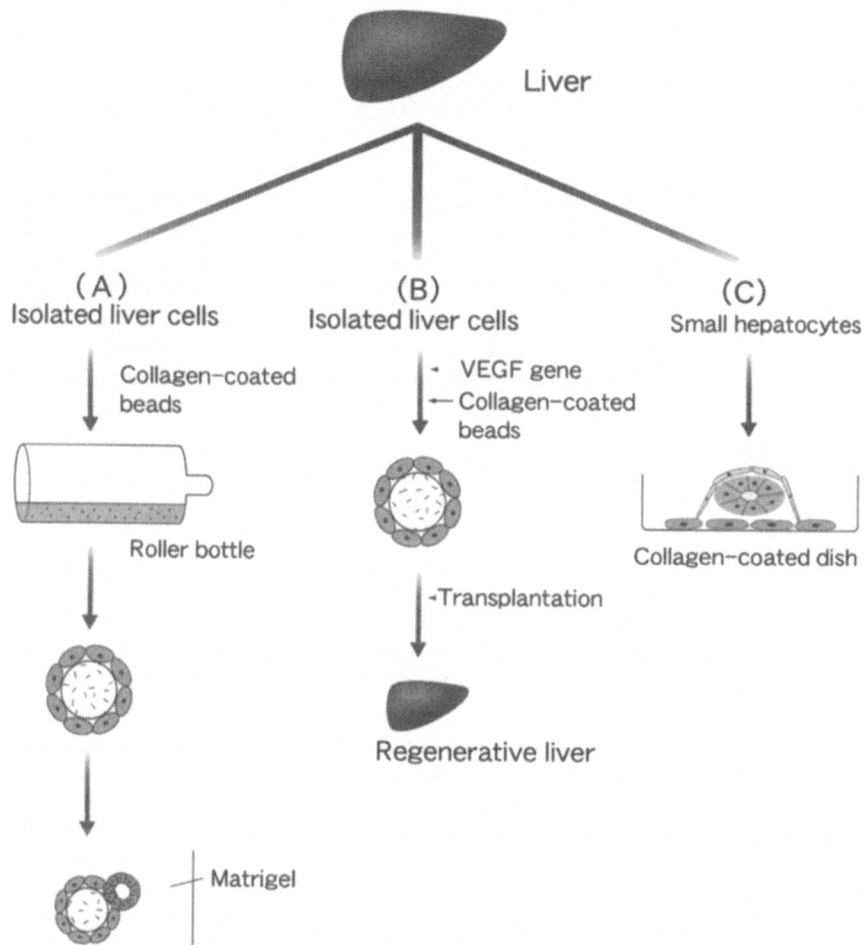

Fig. 1A–C. Schematic diagram illustrating the hypothesis that oval cells develop from stem cells located in biliary ductules and that they differentiate into hepatocytes and bile duct. *VEGF* vascular endothelial growth factor

to cells that can secrete biogenic VEGF proteins. Subsequently, when they cultured transformed hepatocytes in the presence of EGF and HGF, groups of hepatocytes (spheroids) were formed. These spheroids were allowed to adhere to collagen-coated beads, the framework of tissue reconstruction, and were then transplanted into the abdominal cavity of mice. As a result, an albumin-expressing hepatic tissue-like structure with copious blood vessels, measuring approximately 1 cm^3 in cubic volume, was formed (Fig. 1B). However, the formation of bile ducts was not observed in this hepatic tissue-like structure. Therefore, it remains unclear how much this hepatic tissue-like structure can actually compensate for liver functions.

Reconstitution of Hepatic Tissues Using Small Hepatocytes

Small hepatocytes have proliferating activity and can be isolated from a non-parenchymal cell fraction of the liver. Therefore, it is considered that small hepatocytes are the precursor cells of hepatocytes [7,8]. Mitaka et al. [8] disseminated a cell fraction consisting of nonparenchymal cells of the liver into a collagen-coated culture dish, and then cultured this cell fraction in the presence of nicotinic acid amide, ascorbic acid, and dimethyl sulfoxide (DMSO). As a result, they found that small hepatocytes initially formed colonies and then piled up to form tubular or cystic structures after prolonged culturing (Fig. 1C). This tubular tissue had bile capillary-like functions. In addition, proliferated small hepatocytes further differentiated into albumin-secreting cells. Therefore, it was considered that hepatic tissue-like structures were reconstructed in the culture dish. However, the number of proliferated small hepatocytes was limited and insufficient for clinical use. This issue should be further investigated in the future.

Reconstitution of Hepatic Tissues Using Liver Stem Cells

Stem cells are defined as cells with abilities to generate new stem cells as the result of self-reproduction as well as to differentiate into mature cells. Therefore, stem cells are classified roughly into two groups on the basis of their differentiation ability [9]. One group consists of embryonic stem cells (ES cells) that can differentiate to almost all cells constituting the body, and the other group consists of adult stem cells that are present in mature tissues or organs. Although the reconstruction of various tissues has been thoroughly studied using ES cells, there are currently many technical and ethical problems that require solving. In recent years, the presence of various adult stem cells has been demonstrated in various organs [9], including the liver [10]. In normal conditions, although the proliferating activity of adult stem cells is not very high, adult stem cells transiently transform into cells with a higher proliferating activity after stimulation with surrounding microenvironments. Subsequently, adult stem cells differentiate into mature cells required for organ formation, which results in tissue regeneration [9]. Therefore, the use of adult stem cells with both proliferating and differentiating activities may be advantageous for tissue reconstruction.

Differentiation and Transplantation of Liver Stemlike Cells (Oval Cells)

Hepatocytes and biliary epithelia are phenotypically very dissimilar, but share a common ancestry. When hepatocytes are severely damaged, a potential stem cell system of biliary origin is activated to generate new hepatocytes (Fig. 2). Oval cells and hepatic epithelial cells are candidate liver stem cells. Oval cells are groups of undifferentiated cells proliferating around the bile ductules in the portal area when residual hepatocytes cannot proliferate under certain pathological conditions, such as a wide variety of hepatopathies. Moreover, hepatic epithelial cells are cells that can be isolated from a nonparenchymal cell fraction of the healthy rat liver. The morphology and characteristics of hepatic epithelial cells are very similar to those of oval cells.

Fig. 2. Reconstitution of hepatic tissues using the primary culture of isolated hepatocytes or small hepatocytes

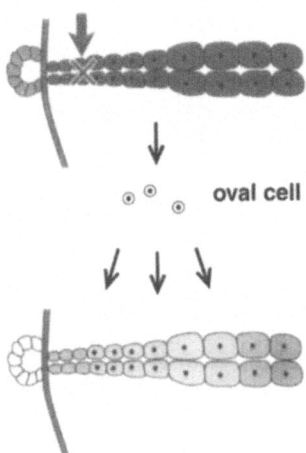

oval cell

It was previously reported that both cells differentiate into hepatocytes or bile epithelial cells in vivo [10,11].

We isolated oval cells from the liver of Long-Evans cinnamon (LEC) rats (11). The cells were γ-glutamyl transpeptidase (γ-GTP) positive, α-fetoprotein positive, cytokeratin-18 and -19 positive, but albumin negative. As the oval cells had dual characteristics of hepatocytes and biliary epithelial cells, we determined whether they are transformed into hepatocytes or bile duct. We infected the oval cells with a LacZ-transducing retrovirus (SG virus) and transplanted them into the liver of NAR/LEC rats. We examined X-gal-stained liver sections from the transplanted NAR/LEC rat and immunohistochemically stained them with anti-albumin antibody. NAR/LEC hepatocytes without the transplantation had negligible immunoreactivity for albumin. On the other hand, several areas in the transplanted liver were heavily stained with anti-albumin antibody, and the X-gal-stained cells were always immunoreactive to albumin. These results indicate that the transplanted oval cells produced albumin as mature hepatocytes and were transformed into hepatocytes (Fig. 3).

We evaluated the efficiency of transplanted oval cells in producing exogenous substances. When transplanted into the liver of NAR/LEC rats, the hepatocytes derived only from the oval cells produced albumin. Transplanting of 4.6×10^6 cells resulted in a 0.3 mg/ml increase in serum albumin. As the normal rat has 35 mg/ml serum albumin, the transformed hepatocytes produced about 1% of the normal level. This result indicated that the transplanted and transformed hepatocytes derived from oval cells produced albumin as efficiently as normal hepatocytes.

Primary culture and genetic manipulation of hepatocytes requires a high degree of skill, and long-term culture of hepatocytes is still impossible. In contrast to hepatocytes, oval cells can be cultured for a long time, genetically manipulated in vitro with ease, and transplanted as hepatocytes in vivo. For ex vivo gene therapy of the liver, oval cells seem better than hepatocytes.

There are many problems that require solving before reconstituting hepatic tissue-like structures with higher-level liver functions using human hepatocytes. However, the development of therapeutic tactics substituting for liver transplantation is an

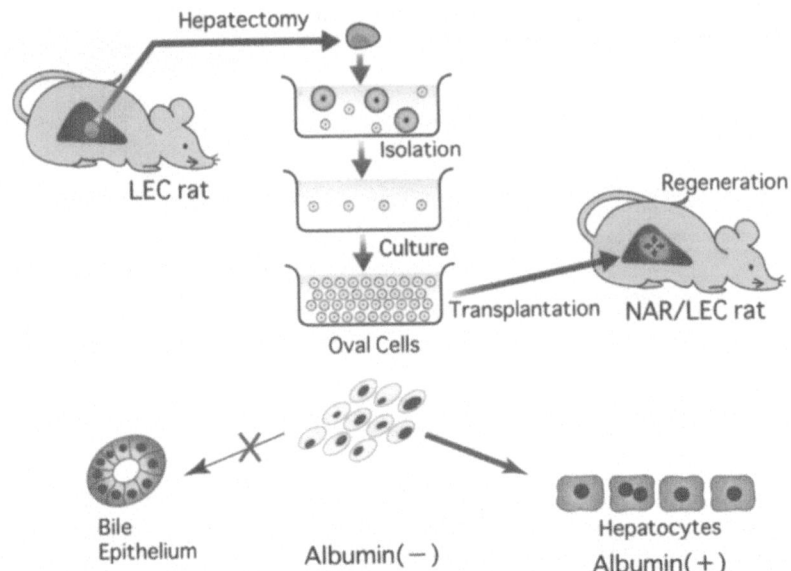

Fig. 3. Differentiation and transplantation of liver stem-like cells (oval cells). When oval cells were transplanted to the liver, they were transformed into hepatocytes

urgent issue. Therefore, the reconstruction of hepatic tissues that can be used in clinical settings will contribute greatly to society.

Acknowledgments. This study was supported in part by Grant-in-Aids for Scientific Research from the Ministry of Education, Science, Sports and Culture of Japan, a grant for "Research for the Future" program from the Japan Society for the Promotion Science (JSPS), and grants from the Naito Foundation and Takeda Science Foundation.

References

1. Strom SC, Fisher RA, Thompson MT, Sanyal AJ, Cole PE, Ham JM, Posner MP (1997) Hepatocyte transplantation as a bridge to orthotopic liver transplantation in terminal liver failure. Transplantation 63:559–569
2. Grossman M, Rader DJ, Muller DW, Kolansky DM, Kozarsky K, Clark BJ 3rd, Stein EA, Lupien PJ, Brewer HB Jr, Raper SE, Wilson JM (1995) A pilot study of ex vivo gene therapy for homozygous familial hypercholesterolaemia. Nat Med 11:1148–1154
3. Fox IJ, Chowdhury JR, Kaufman SS, Goertzen TC, Chowdhury NR, Warkentin PI, Dorko K, Sauter BV, Strom SC (1998) Treatment of the Crigler-Najjar syndrome type I with hepatocyte transplantation. N Engl J Med 338:1422–1426
4. Chen SC, Mullon C, Kahaku E, Watanabe F, Hewitt W, Eguchi S, Middleton Y, Arkadopoulos N, Rozga J, Solomon B, Demetriou AA (1997) Treatment of severe liver failure with a bioartificial liver. Ann NY Acad Sci 831:350–360
5. Michalopoulos GK, Bowen WC, Zajac VF, Beer-Stolz D, Watkins S, Kostrubsky V, Strom SC (1999) Morphogenetic events in mixed cultures of rat hepatocytes and non-

parenchymal cells maintained in biological matrices in the presence of hepatocyte growth factor and epidermal growth factor. Hepatology 29:90–100

6. Ajioka I, Akaike T, Watanabe Y (1999) Expression of vascular endothelial growth factor promotes colonization, vascularization, and growth of transplanted hepatic tissues in the mouse. Hepatology 29:396–402

7. Tateno C, Yoshizato K (1996) Growth and differentiation in culture of clonogenic hepatocytes that express both phenotypes of hepatocytes and biliary epithelial cells. Am J Pathol 149:1593–1605

8. Mitaka T, Sato F, Mizuguchi T, Yokono T, Mochizuki Y (1999) Reconstruction of hepatic organoid by rat small hepatocytes and hepatic nonparenchymal cells. Hepatology 29:111–125

9. Fuchs E, Serge JA (1999) Stem cells: a new lease on life. Cell 100:143–155

10. Grisham JW, Thorgeisson SS (1997) Liver stem cells. In: Potten CS (ed) Stem cells. Academic Press, London, pp 233–282

11. Yasui O, Miura N, Terada K, Kawarada Y, Koyama K, Sugiyama T (1997) Isolation of oval cells from Long-Evans cinnamon rats and their transformation into hepatocytes in vivo in the rat liver. Hepatology 25:329–334

Regulation of TNF-α- and Fas-Induced Hepatic Apoptosis by NF-κB

David A. Brenner, Etsuro Hatano, Cynthia Bradham, Robert Schwabe, Yuji Iimuro, Ting Qian, Ronald Thurman, and John Lemasters

Summary. Tumor necrosis factor (TNF)-α and Fas are two potent hepatotoxic agonists in human diseases and animal models. When either TNF-α or Fas ligand (FasL) binds to its respective receptor, it initiates parallel cascades leading to either apoptosis or the induction of antiapoptotic proteins. The transcription factor NF-κB is a critical mediator of antiapoptosis in the liver in vivo and in cultured hepatocytes. The mechanisms by which NF-κB blocks the apoptotic pathway are under intense investigation. In the hepatocyte, NF-κB induces at least two antiapoptotic proteins, inducible nitric oxide synthase (iNOS) and Bcl-xL. Understanding the regulation of NF-κB and apoptosis in hepatocytes may provide insights into the development of novel therapies for liver disease.

Key words. Hepatocyte, Apoptosis, fas, Mitochondria, Tumor necrosis factor

Introduction

The transcription factor NF-κB is normally located in the cytoplasm in an inactive state bound to its inhibitor, IκB. Activation of an upstream kinase complex, IKK, phosphorylates the IκB at serines 32 and 36, leading to their subsequent ubiquination and degradation by proteosomes. The IKK complex contains at least three components, the kinases IKK-1 and IKK-2, as well as the scaffold-type protein IKK AP (also called Nemo or IKK-γ). The upstream activating kinase for IKK is still unresolved, but it has been proposed to include NIK and MEKK-1. It is through this unresolved upstream kinase (MAKKK) that external agonists such as tumor necrosis factor (TNF)-α binding to their cognate receptors lead to downstream activation of NF-κB [1–5] (Fig. 1).

Previous studies from our laboratory and others have demonstrated that NF-κB is activated by models of liver regeneration and liver transplantation. However, the function of NF-κB activation in the liver was unknown. Was the activation a critical component of hepatocyte physiology, or did it merely represent an artifact of surgical intervention? More recent studies have demonstrated that the NF-κB p65 knockout mouse and the IKK-2 knockout mouse are embryonic lethal with massive hepatic

Campus Box 7038, 156 Glaxo Building, University of North Carolina at Chapel Hill, Chapel Hill, NC 27599-7038, USA

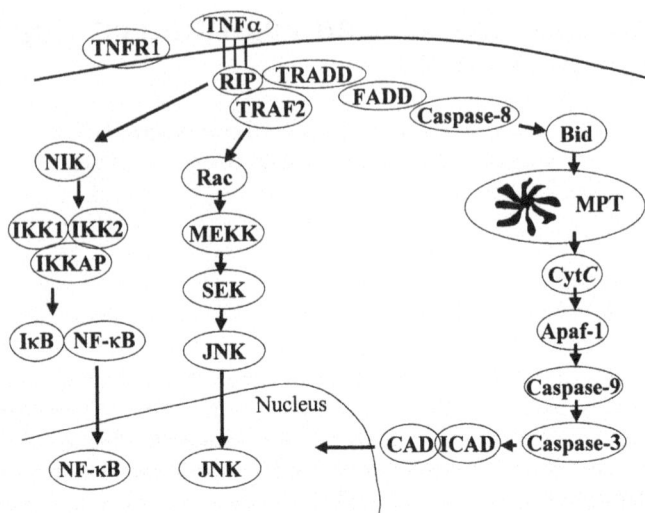

Fig. 1. Model for the signaling pathways of tumor necrosis factor-alpha (*TNF-α*). After TNF-α binds to TNFR₁, the pathways may induce NF-κB, JNK, or apoptosis

apoptosis [6,7]. The knockout mice provided corroborative support for the role of NF-κB in hepatocytes but prevented the use of these mice for studies in hepatic pathophysiology in adult mice.

NF-κB Prevents Apoptosis in Liver Regeneration and in Liver Transplantation

As an alternative method to the use of conventional knockout mice, we developed an adenovirus that expresses an IκB superrepressor (Fig. 2). The phosphorylation targets serine 32 and serine 36 are mutated to alanines, which prevents IκB phosphorylation and subsequent degradation. Thus, the IκB superrepressor binds irreversibly to the NF-κB and prevents NF-κB activation. In the rat model of liver regeneration following partial hepatectomy, there is a marked proliferative response 24 h after partial hepatectomy, which is documented by DNA synthesis or by the mitotic index. On the other hand, if the rats are pretreated with the adenovirus containing the IκB superrepressor, then there is massive apoptosis instead of proliferation at 24 h following partial hepatectomy. Therefore, blocking NF-κB converts a proliferative response to an apoptotic response in the liver in vivo [8].

We also used a model of liver transplantation in rats in which the donor liver was stored for 24 h in University of Wisconsin Solution and then transplanted into a recipient. Warm reperfusion, but not cold ischemia, was a strong inducer of NF-κB DNA-binding activity [9]. Pretreating the donor with the IκB superrepressor produced massive hepatic apoptosis and congestion after liver transplantation. A control adenovirus did not produce similar damage [10]. Thus, NF-κB activation is required

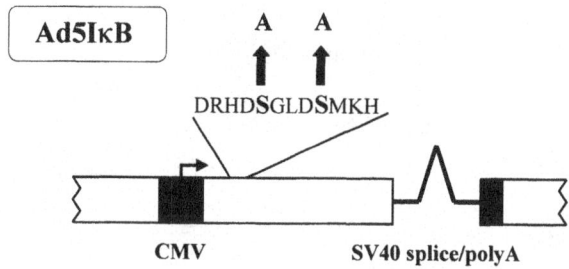

Fig. 2. Recombinant adenovirus expressing IκB superrepressor

to prevent apoptosis in the model of cold ischemia/warm reperfusion of liver transplantation.

The Mitochondrial Permeability Transition Is Required for TNF-α-Mediated Apoptosis

Previous studies demonstrated that TNF-α is required for liver regeneration using either anti-TNF-α antibody or TNFR1 knockout mice. TNF-α is a strong inducer of NF-κB, including in the liver. TNF-α is associated with liver injury in experimental alcoholic liver disease and in liver transplantation and is elevated in patients with chronic hepatitis C viral infection and alcoholic liver disease [11].

To assess the role of TNF-α in apoptosis in primary hepatocytes, we pretreated hepatocytes with the IκB superrepressor and then assessed the role of TNF-α. Blocking NF-κB with the IκB superrepressor resulted in TNF-α-mediated apoptosis in primary rat and mouse hepatocytes [12,13]. Apoptosis was assessed by cellular morphology, nuclear fragmentation revealed by DAPI staining, DNA fragmentation revealed by agarose gel electrophoresis, and activation of the executionary caspase 3. The role of the mitochondrial permeability transition (MPT) was assessed in primary hepatocytes using confocal microscopy. The MPT is a high-conductance, nonspecific pore that conducts solutes as large as 1500 daltons. The MPT pore is apparently formed by the adenosine nucleotide translocator, cyclophilin D, and the voltage-dependent anion channel. The MPT causes mitochondrial depolarization, uncoupling, and swelling. MPT was assessed by simultaneously monitoring the anionic fluorophor TRM and the neutral fluorophor calcein. The induction of the MPT results in the loss of TMRM from the mitochondria and the loss of exclusion of calcein from the mito-chondria. These studies demonstrated that TNF-α induces the MPT and mitochon-drial depolarization only when NF-κB is blocked in hepatocytes. The MPT is critical for apoptosis induced by TNF-α in that cyclosporin A will block the MPT and block apoptosis. Using a variety of inhibitors, we demonstrated that the MPT is downstream from the adaptor protein FADD and the initiator caspase 8, but it is upstream from the release of cytochrome *c* into the cytoplasm and activation of the executionary caspase 3 (Fig. 3).

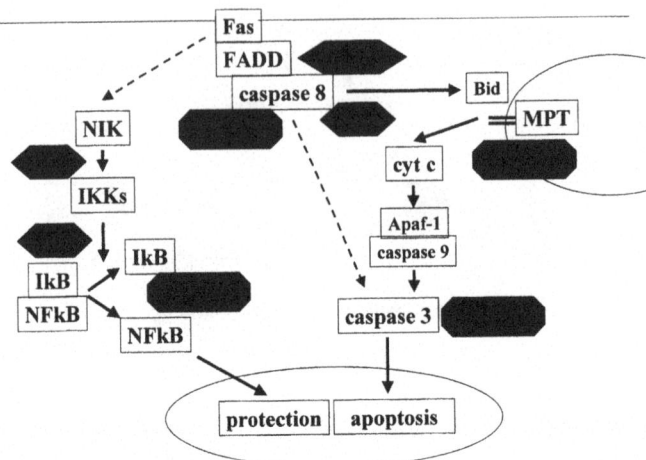

Fig. 3. Model for the signaling pathway of Fas. Fas may induce the mitochondrial permeabitity transition (*MPT*) or may activate caspase 8 to directly activate caspase 3

The MPT Augments Fas-Induced Apoptosis in Mouse Hepatocytes

We next wished to assess the signaling pathway by which Fas induces apoptosis in mouse hepatocytes. Fas is expressed on mouse and human hepatocytes. Activating Fas by Fas ligand or anti-Fas antibodies produces fulminant hepatitis in mice. Fas plus actinomycin D induces apoptosis in cultured mouse hepatocytes. Fas mediates apoptosis in models of cholestasis in the liver, and Fas is overexpressed in chronic viral hepatitis.

We demonstrated that pretreatment with the IκB superrepressor sensitizes mouse hepatocytes to TNF-α and Fas-mediated apoptosis. Alternatively, blocking NF-κB activation with a proteosome inhibitor also induces TNF-α- and Fas-mediated apoptosis. Both TNF-α and the Fas agonistic antibody Jo2 activate caspase 3 and caspase 8 in IκB-sensitized hepatocytes. However, Jo2 induced an early and strong activation of caspase 8 not seen with TNF-α treatment. Additional experiments using adenoviruses expressing dominant-negative IKK-2 and dominant-negative IKK-1 demonstrated that IKK-2 was the critical kinase in NF-κB induction. Blocking IKK-2 sensitized hepatocytes to TNF-α- or Jo2-induced apoptosis [13] (Schwabe et al., in manuscript).

Similar to TNF-α, Fas induced the MPT in IκB superrepressor-sensitized hepatocytes. The MPT was blocked by pretreating mouse hepatocytes with cyclosporin A and trifluoroperizine. However, surprisingly, blocking the MPT failed to protect the hepatocytes from Jo2-induced apoptosis in contradistinction to the protection from TNF-α-induced apoptosis. Confocal microscopy revealed that Fas can still induce apoptosis even in the presence of polarized mitochondria. A time-course study revealed that blocking the MPT delayed but did not prevent killing by Jo2. Blocking the MPT also prevented the release of cytochrome *c* from the mitochondria by Jo2.

We propose the following model (see Fig. 3) for Fas-induced apoptosis in hepatocytes. Fas can induce the MPT, which in turn causes the release of cytochrome c and the activation of executionary caspases. However, when the MPT is blocked, then the strong induction of caspase 8 is sufficient, with a slower time course, to induce caspase 3 directly and still induce apoptosis.

iNOS Is an NF-κB-Responsive Gene That Blocks Apoptosis in Hepatocytes

Previous studies have demonstrated that inducible nitric oxide synthase (iNOS) is an NF-κB-responsive gene induced by TNF-α. NO, the product of iNOS, has been implicated as a protective factor or as an apoptotic factor in different cell types. We speculate that iNOS may be one NF-κB-responsive gene that protects from TNF-α-mediated apoptosis in hepatocytes.

We demonstrated that the NO donor, SNAP, inhibits TNF-α- and Fas-mediated apoptosis in IκB-sensitized mouse hepatocytes. Thus, pharmacological doses of NO are sufficient to prevent apoptosis. On the other hand, the NOS inhibitor N^G-monomethyl-L-arginine (L-NMMA) sensitizes mouse hepatocytes to TNF-α- and Fas-mediated apoptosis even in the absence of blocking NF-κB. To use a genetic approach to assess the role of iNOS in apoptosis, we used iNOS knockout mice. These mice developed normally and have a normal phenotype. We demonstrated that TNF-α or Jo2 induces iNOS in wild-type mice but not in iNOS knockout mice. Blocking NF-κB with the IκB superrepressor prevented the induction of iNOS in wild-type hepatocytes. Finally, we demonstrated that iNOS knockout hepatocytes were sensitive to TNF-α- and Fas-mediated apoptosis, even in the absence of the IκB superrepressor (Hatano et al., in manuscript). Thus, one of the NF-κB-responsive genes that prevents apoptosis by TNF-α- or Fas-mediated apoptosis is iNOS.

Fas Induces the Antiapoptotic Kinase, AKT

AKT is an antiapoptotic kinase that mediates its effects through the phosphorylation of BAD, the phosphorylation of caspase 9, and the activation of the forkhead transcription factors. Most recently, AKT was demonstrated to activate NF-κB either through activation of its upstream kinase IKK or by a more direct activation of NF-κB transcriptional activity. We demonstrated that Jo2 and, to a lesser extent, TNF-α both phosphorylate and activate AKT. Inhibiting the upstream kinase PI-3 kinase sensitized mouse hepatocytes to apoptosis by TNF-α or Jo2, even in the absence of the IκB superrepressor. The PI-3 kinase inhibitor LY294002 blocked the activation of NF-κB by TNF-α or Jo2, implying that AKT was required for complete activation of NF-κB by these two agonists. The critical role of AKT in the signal transduction pathway was demonstrated by using a constitutively activated AKT delivered by adenovirus. Constitutively activated AKT rescued LY294002-sensitized hepatocytes from TNF-α- or Jo2-induced apoptosis. Thus, AKT is the critical downstream target of PI-3 kinase and is a key antiapoptotic mediator via its activation of NF-κB in hepatocytes.

We demonstrated that both TNF-α and Fas induce Bcl-xL expression in hepatocytes. This expression is blocked by pretreatment with either the IκB superrepressor or the PI-3 kinase inhibitor. Thus, the critical antiapoptotic protein induced by AKT and NF-κB by Fas is Bcl-xL. Finally, using hepatocytes derived from a mouse that constitutively overexpresses Bcl-xL in hepatocytes demonstrated that Bcl-xL protected from Jo2-induced apoptosis in hepatocytes sensitized with either actinomycin D or IκB superrepressor.

References

1. Malinin NL, Boldin MP, Kovalenko AV, Wallach D (1997) MAP3K-related kinase involved in NF-kappaB induction by TNF, CD95 and IL-1. Nature 385(6616):540–544
2. Natoli G, Costanzo A, Moretti F, Fulco M, Balsano C, Levrero M (1997) Tumor necrosis factor (TNF) receptor 1 signaling downstream of TNF receptor-associated factor 2. Nuclear factor kappaB (NFkappaB)-inducing kinase requirement for activation of activating protein 1 and NFkappaB but not of c-Jun N-terminal kinase/stress-activated protein kinase. J Biol Chem 272(42):26079–26082
3. DiDonato JA, Hayakawa M, Rothwarf DM, Zandi E, Karin M (1997) A cytokine-responsive IkappaB kinase that activates the transcription factor NF-kappaB. Nature 388(6642):548–554
4. Mercurio F, Zhu H, Murray BW, Shevchenko A, Bennett BL, Li J, Young DB, Barbosa M, Mann M, Manning A, Roa A (1997) IKK-1 and IKK-2: cytokine-activated IkappaB kinases essential for NF-kappaB activation. Science 278(5339):860–866
5. Zandi E, Rothwarf DM, Delhase M, Hayakawa M, Karin M (1997) The IkappaB kinase complex (IKK) contains two kinase subunits, IKKalpha and IKKbeta, necessary for IkappaB phosphorylation and NF-kappaB activation. Cell 91(2):243–252
6. Beg AA, Sha WC, Bronson RT, Ghosh S, Baltimore D (1995) Embryonic lethality and liver degeneration in mice lacking the RelA component of NF-kappa B. Nature 376(6536):167–170
7. Li Q, Van Antwerp D, Mercurio F, Lee KF, Verma IM (1999) Severe liver degeneration in mice lacking the IkappaB kinase 2 gene. Science 284(5412):321–325
8. Iimuro Y, Nishiura T, Hellerbrand C, Behrns KE, Schoonhoven R, Grisham JW, et al (1998) NFkappaB prevents apoptosis and liver dysfunction during liver regeneration [published erratum appears in J Clin Invest 1998 Apr 1;101(7):1541]. J Clin Invest 101(4):802–811
9. Bradham CA, Stachlewitz RF, Gao W, Qian T , Jayadev S, Jenkins G, et al (1997) Reperfusion after liver transplantation in rats differentially activates the mitogen-activated protein kinases. Hepatology 25(5):1128–1135
10. Bradham CA, Schemmer P, Stachlewitz RF, Thurman RG, Brenner DA (1999) Activation of nuclear factor-kappaB during orthotopic liver transplantation in rats is protective and does not require Kupffer cells. Liver Transplant Surg 5(4):282–293
11. Brenner DA (1999) Therapeutic strategy for liver fibrosis. J Gastroenterol Hepatol 14S: A279–A280
12. Bradham CA, Qian T, Streetz K, Trautwein C, Brenner DA, Lemasters JJ (1998) The mitochondrial permeability transition is required for tumor necrosis factor alpha-mediated apoptosis and cytochrome *c* release. Mol Cell Biol 18(11):6353–6364
13. Hatano E, Bradham CA, Stark A, Iimuro Y, Lemasters JJ, Brenner DA (2000) The mitochondrial permeability transition augments Fas-induced apoptosis in mouse hepatocytes. J Biol Chem 275(16):11814–11823

Retinoid-Induced Apoptosis in Hepatocellular Carcinoma: A Molecular Basis for "Clonal Deletion" Therapy

Masataka Okuno[1], Hisataka Moriwaki[1], Rie Matsushima-Nishiwaki[1], Tetsuro Sano[1], Seiji Adachi[1], Kuniharu Akita[1], and Soichi Kojima[2]

Summary. We propose a new concept of "clonal deletion" therapy for chemoprevention of hepatocellular carcinoma (HCC). We previously showed that administration with acyclic retinoid suppressed alpha-fetoprotein (AFP)-L3-positivity and subsequently reduced the incidence of a second HCC in cirrhotic patients who had undergone curative treatment for previous liver cancers. Such eradication of AFP-L3-producing latent malignant (or premalignant) cells from the liver suggested a new strategy to prevent HCC, which may be identified with cancer chemotherapy. In the present study, we explored the molecular mechanism of clonal deletion and found a novel mechanism of apoptosis induction by retinoid. We demonstrated a modification of a retinoid receptor, RXR-α, by mitogen-activated protein kinase-dependent phosphorylation, resulting in the loss of transactivating activity. This change may lead HCC cells to be resistant to natural retinoic acid. However, acyclic retinoid restored the function of phosphorylated RXR-α and induced its downstream gene, including tissue transglutaminase, an enzyme that has been implicated in apoptosis. Tissue transglutaminase-dependent apoptosis in HCC cells was independent of the activation of caspases. This novel mechanism of retinoid-induced apoptosis may provide a clue for understanding the molecular mechanism of clonal deletion.

Key words. Retinoid, Cancer chemoprevention, Clonal deletion, Apoptosis, Retinoid receptor, Tissue transglutaminase

Introduction

Chemoprevention of hepatocellular carcinoma (HCC) is of great significance because of the high incidence of HCC in viral-associated cirrhotic patients. The annual incidence of HCC reaches 5% to 7% in both type B and type C hepatitis virus-infected cirrhotic patients. Moreover, the incidence increases up to approximately 20% per year in cirrhotic patients who have undergone curative treatment of the primary HCC. This high incidence of second HCC, half of which is a second primary cancer and not an intrahepatic metastatic tumor, is a major cause of the low 5-year survival rate (40%)

---

[1] First Department of Internal Medicine, Gifu University School of Medicine, 40 Tsukasa-machi, Gifu 500-8705, Japan
[2] Laboratory of Molecular Cell Sciences, Tsukuba Institute, The RIKEN, Tsukuba, Japan

after curative treatment of HCC [1]. Therefore, future advances in merely medical technologies in both the early detection and the therapy of HCC may not achieve further improvement of the therapeutic outcome of HCC; a new strategy to prevent second primary HCC may be required for this purpose.

Some strategies have been reported to suppress hepatocarcinogenesis in clinical trials, including interferon-α and -β [2], a Japanese herbal medicine, TJ-9 [3], and our retinoid analogue, acyclic retinoid [4,5]. Both interferon and TJ-9 may belong to the category of immunopreventive agents, whereas acyclic retinoid is chemopreventive. At present, interferon and TJ-9 have been shown to prevent the occurrence of primary HCC in cirrhotic patients, and acyclic retinoid suppressed the appearance of a second HCC after curative treatment of preceding tumors.

Clonal Deletion by Acyclic Retinoid

Acyclic retinoid (all *trans*-3,7,11,15-tetramethyl-2,4,6,10,14-hexadecapentanoic acid), or NIK 333, inhibited experimental liver carcinogenesis [5] and induced the apoptosis of human hepatoma-derived cell lines [6]. We demonstrated that oral administration of acyclic retinoid for 12 months significantly reduced the incidence of second primary cancers in patients who had received curative treatment of hepatocellular carcinomas in a randomized controlled trial [7]. Moreover, the survival rate was also significantly improved by the compound after a median observation period of 62 months in the follow-up study [4]. In that clinical trial, serum lectin-reactive α-fetoprotein (AFP-L3), which indicates the presence of immature hepatic cells in the remnant liver, disappeared in the acyclic retinoid group after a 12-month administration. This observation suggests AFP-L3-producing clones were eradicated by acyclic retinoid from the remnant liver [8]. In contrast, spontaneous reduction in serum AFP-L3 levels was not observed in the placebo group. Thus, we have suggested that acyclic retinoid deleted premalignant or latent malignant clones that produce AFP-L3 from the remnant liver. Moreover, acyclic retinoid suppressed the appearance of serum AFP-L3 in patients whose AFP-L3 levels were below the detection range, whereas increased numbers of patients whose serum levels of AFP-L3 were newly enhanced were detected in the placebo group. From these results, we have proposed the new concept of cancer chemoprevention, "clonal deletion" and "clonal inhibition" (Fig. 1) [9]. These ideas may lead us to identify chemoprevention with chemotherapy. Because of the high incidence of HCC in cirrhotic patients, it may well be possible some latent malignant (or premalignant) cells are already present in the cirrhotic liver that cannot be detected by diagnostic image analyses. Acyclic retinoid might remove such clones from the liver and thereby contribute to reduce the incidence of a second HCC. This concept may provide a good reason to place prevention and therapy in the same category.

Retinoid Resistance in Hepatocarcinogenesis

We have shown a depletion of retinoid in tumorous tissues in both experimental and clinical HCC [5]. The absence of hepatic stellate cells, retinoid-storing cells in the liver, is a major cause of such retinoid depletion in the tumors. In addition, the rapid

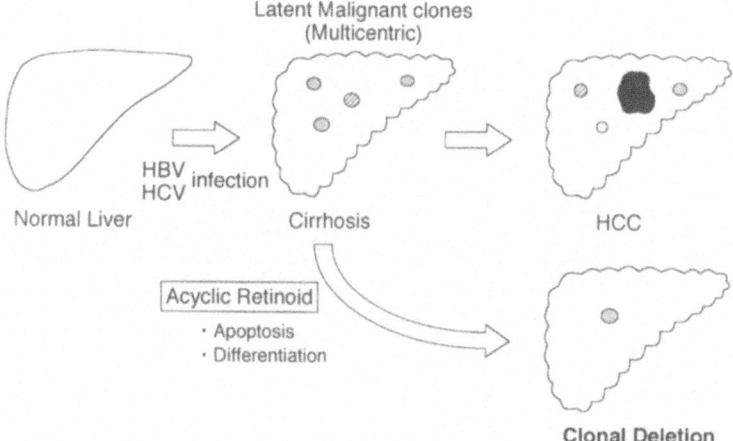

Fig. 1. The concept of "clonal deletion" therapy. A high incidence of hepatocellular carcinoma (*HCC*) in cirrhotic patients strongly suggests the presence of minute or latent malignant cells in the liver that cannot be diagnosed clinically by image analyses. Therefore, eradication of such transformed clones (clonal deletion) may be recognized as a therapy instead of prevention. Our clinical experience suggests deletion of alpha-fetoprotein-L3-producing transformed clones from the liver by acyclic retinoid therapy. Once such clones are deleted, the preventive effect on HCC lasts for years without continued administration of the retinoid

metabolism of retinoid into an inactive metabolite may well participate in such depletion [5]. Retinol has been shown to be rapidly converted to anhydroretinol, an inactive retinol metabolite that cannot be metabolized to retinoic acid, a ligand of the retinoid nuclear receptor. This retinoid depletion might be involved in hepatocarcinogenesis. For instance, we have suggested that the enhanced loss of retinoid in the liver in cirrhotics with viral infection plus alcohol abuse may be linked to the accelerated incidence of HCC [10]. Because there have been several reports showing chemopreventive effects of retinoids on liver cancer in both experimental and clinical studies, it is conceivable that retinoid deficiency may promote tumorigenesis in the liver.

In addition to retinoid depletion, we have also found that a malfunction of retinoid nuclear receptor in hepatoma cells may contribute to hepatocarcinogenesis. Retinoids exert their biological effects through two distinct nuclear receptors, retinoic acid receptor (RAR) and retinoid X receptor (RXR) [11]. RAR interacts with both all-*trans* retinoic acid (atRA) and 9-*cis* retinoic acid (9cRA), whereas RXR binds only to 9cRA and regulates the expression of several target genes. Both RAR and RXR consist of three subtypes, α, β, and γ, characterized by a modular domain structure. RXR forms a homodimer as well as heterodimers with RAR and other nuclear receptors. These dimers bind to their respective response elements and subsequently activate or inhibit the expression of their target genes. RAR and RXR bind to a retinoic acid response element (RARE) and RXR response element (RXRE), respectively. It is suggested that both RAR and RXR are involved in hepatocarcinogenesis. For example, expression of the RAR-α gene is localized near the integration site of hepatitis B virus and is induced

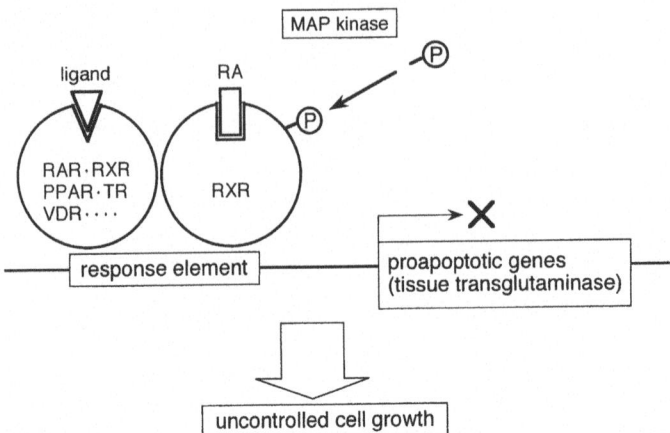

Fig. 2. Retinoid-refractory HCC cells. RXRα, a retinoid nuclear receptor, is posttranslationally modified by phosphorylation in cancer cells, leading to a loss of its transactivating activity. Failure in inducing its downstream genes that regulate cellular proliferation and apoptosis, such as tissue transglutaminase, may lead the cells to uncontrolled growth. *MAP*, mitogen-activated protein

in HCC tissue [12]. On the other hand, RXR-α is reported to bind to the enhancer element of hepatitis B virus and modulate viral replication [13].

Among these receptors, RXR-α is most abundant in the liver and is highly expressed in HCC cells. We have shown a delay in the metabolism of RXR-α that remained a full-length 54-kDa protein, whereas RXR-α rapidly decayed into smaller peptides in the surrounding cirrhotic and normal liver tissues. Because there was no genomic mutation in the RXR-α gene in either surgically resected human HCC tissues or human HCC cell lines (Adachi et al., unpublished data), we hypothesized that post-translational modification of RXR-α may take place and be responsible for the delayed proteolytic decay of the receptor. Many members of the nuclear hormone receptor family are phosphoproteins [14]. Recently, phosphorylation of RXR-α by mitogen-activated protein (MAP) kinase has been reported in keratinocytes, which may play an important role in their malignant transformation [15].

We have reported that the nonlysosomal calcium-dependent cysteine protease, *m*-calpain, participates in the proteolysis of RXR-α [16]. Some substrates of *m*-calpain, such as c-Fos, may become resistant to proteolysis after their phosphorylation [17]. Very recently, we have found that full-length RXR-α protein was phosphorylated, but that this was not the case with proteolytically truncated RXR-α protein in both surgically resected tissues and HCC cell lines [18] (Fig. 2). Phosphorylation took place at some serine and threonine residues, consensus sites of MAP kinases that played a significant role in such phosphorylation. Very interestingly, phosphorylated RXR-α lost its transactivating activity, which seems to correlate with enhanced proliferation of HCC cells. Although we have not yet fully identified the target gene(s) of RAR-α that regulate cellular proliferation, these observations may suggest that malfunction of phosphorylated RXR-α is linked to the aberrant growth of HCC. Therefore, in HCC tissues and cells, not only retinoid depletion but also malfunction of modified retinoid nuclear receptor may be involved in the development of HCC.

Induction of Apoptosis by Acyclic Retinoid

Clonal deletion therapy suggests a removal of latent malignant (or premalignant) cells from the liver. In general, two possible mechanisms are widely accepted for such removal of transformed cells: cell death (or apoptosis) and differentiation induction. These two mechanisms may work independently in some cases but coincide in many systems. One typical example is retinoic acid therapy against acute promyelocyte leukemia (APL) [19]. Retinoic acid induces the differentiation of APL cells into granulocytes. Once differentiated, granulocytes restore their normal function and terminate their lives by apoptosis. This phenomenon is recognized as terminal differentiation. Similar mechanisms are hypothesized also in the chemoprevention of several cancers.

In the case of HCC, we have shown the induction of both apoptosis [6] and differentiation induction [20]. At present, differentiation of HCC cells by acyclic retinoid requires further investigation because we have shown only the upregulation of albumin with downregulation of α-fetoprotein [20] but failed to show the phenotypic differentiation of HCC cells. In contrast, apoptosis induction of HCC cells by acyclic retinoid has been shown clearly, and its molecular mechanism(s) has been suggested. Here, immediate apoptosis that was not accompanied by differentiation has been studied. Kojima et al. have proposed an involvement of tissue transglutaminase in the retinoic acid-induced apoptosis in APL cells, which may take place independently of the activation of caspases (Kojima et al., unpublished observation).

There have been several reports showing an induction of tissue transglutaminase in retinoic acid-induced apoptotic cells [21]. However, the function of tissue transglutaminase is not yet understood. We found that acyclic retinoid-induced apoptosis in the HCC cell line HuH7 cells was also dependent on tissue transglutaminase (Sano, unpublished data). This apoptosis was not inhibited by caspase inhibitors but was suppressed by antisense oligo DNA specific to tissue transglutaminase. HuH7 cells treated with acyclic retinoid showed typical features of apoptotic changes including chromatin condensation when examined by electron microscopy as well as DNA ladder formation. We therefore speculate that both tissue transglutaminase and caspase work simultaneously; the former might participate in chromatin condensation whereas the latter works for DNA fragmentation.

Tissue transglutaminase is one of the downstream genes regulated by the RXR/RXR system. Thus, it might well be possible that malfunction of phosphorylated RXR results in the failure of tissue transglutaminase induction in HCC cells, leading to the uncontrolled proliferation resistant to natural retinoic acid. However, acyclic retinoid might somehow restore the function of RXR and thereby induce apoptosis via tissue transglutaminase pathway. The mechanism by which acyclic retinoid recovers phosphorylated RXR function is now under investigation.

Clinical Application of Clonal Deletion

Clonal deletion therapy has already been theoretically proposed as a mechanism of cancer chemoprevention [22–24] and was strongly supported by our clinical experience of HCC [9]. The present study may provide information to understand the molecular mechanism of clonal deletion. The benefit of clinical application of this new concept may place chemoprevention in the same category with chemotherapy, which will make prevention more acceptable in the clinical field. It is widely recognized that there may be minute or latent malignant cells in the cirrhotic liver in such patients as AFP-L3-positive ones. Therefore, even when the cells cannot be detected by diagnostic imaging, there seems to be good reasons to treat such patients with clonal deletion therapy.

Acknowledgment. This work was supported in part by Grants-in-Aid from the Ministry of Education, Culture, Sports, Science and Technology (M. O., H. M., and S. K.) and from the Ministry of Health, Labour and Welfare of Japan (H. M.).

References

1. Liver Cancer Study Group of Japan (1997) Survey and follow-up study of primary liver cancer in Japan. Acta Hepatol Jpn 38:317–330
2. Nishiguchi S, Kuroki T, Nakatani S, Morimoto H, Taketa T, Nakajima S, Shimi S, Seki S, Kobayashi K, Otani S (1995) Randomized trial of effects of interferon-α on the incidence of hepatocellular carcinoma in chronic active hepatitis C with cirrhosis. Lancet 346:1051–1055
3. Oka H, Yamamoto S, Kuroki T, Harihara S, Marumo T, Kim SR, Monna T, Kobayashi K, Tango T (1995) Prospective study of chemoprevention of hepatocellular carcinoma with Sho-saiko-to (TJ-9). Cancer (Phila) 76:743–749
4. Muto Y, Moriwaki H, Saito A (1999) Prevention of second primary tumors by an acyclic retinoid in patients with hepatocellular carcinoma. N Engl J Med 340:1046–1047
5. Muto Y, Moriwaki H (1984) Antitumor activities of vitamin A and its derivatives. J Natl Cancer Inst 73:1389–1393
6. Nakamura N, Shidoji Y, Yamada Y, Hatakeyama H, Moriwaki H, Muto Y (1995) Induction of apoptosis by acyclic retinoid in the human hepatoma-derived cell line, HuH-7. Biochem Biophys Res Commun 207:382–388
7. Muto Y, Moriwaki H, Ninomiya M, Adachi S, Saito A, Takasaki KT, Tanaka T, Tsurumi K, Okuno M, Tomita E, Nakamura T, Kojima T (1996) Prevention of second primary tumors by an acyclic retinoid, polyprenoic acid, in patients with hepatocellular carcinoma. N Engl J Med 334:1561–1567
8. Moriwaki H, Yasuda I, Shiratori Y, Uematsu T, Okuno M, Muto Y (1997) Deletion of serum lectin-reactive α-fetoprotein by acyclic retinoid: a potent biomarker in the chemoprevention of second primary hepatoma. Clin Cancer Res 3:727–731
9. Moriwaki H, Okuno M, Shiratori Y, Yasuda I, Muto Y (2000) Clonal deletion, a novel strategy of cancer control that falls between cancer chemoprevention and cancer chemotherapy: a clinical experience in liver cancer. In: Okita K (ed) Frontiers in hepatology. Progress in hepatocellular carcinoma treatment. Springer-Verlag, Tokyo, pp 97–103
10. Adachi S, Moriwaki H, Muto Y, Yamada Y, Fukutomi Y, Shimazaki M, Okuno M, Ninomiya M (1991) Reduced retinoid content in hepatocellular carcinoma with special reference to alcohol consumption. Hepatology 14:776–780

11. Mangelsdorf DJ, Umesono K, Evans RM (1994) The retinoids receptors. In: Sprn MB, Roberts AB, Goodman DS (eds) The reinoids: biology, chemistry, and medicine, 2nd edn. Raven, New York, pp 319–349

12. Benbrook D, Lernhardt E, Pfahl M (1988) A new retinoic acid receptor identified from a hepatocellular carcinoma. Nature (Lond) 333:669–672

13. Garcia AD, Ostapchuk P, Hearing P (1993) Functional interaction of nuclear factors EF-C, HNF-4, and RXR α with hepatitis B virus enhancer 1. J Virol 67:3940–3950

14. Weigel N (1996) Steroid hormone receptors and their regulation by phosphorylation. Biochem J 319:657–667

15. Solomon C, White JH, Kremer R (1999) Mitogen-activated protein kinase inhibits 1, 25-dihydroxyvitamin D3-dependent signal transduction by phosphorylating human retinoid X receptor-α. J Clin Invest 103:1729–1735

16. Matsushima-Nishiwaki R, Shidoji Y, Nishiwaki S, Moriwaki H, Muto Y (1996) Limited degradation of retinoid X receptor by calpain. Biochem Biophys Res Commun 225: 946–951

17. Okazaki K, Sagata N (1995) The Mos/MAP kinase pathway stabilizes c-Fos by phosphorylation and augments its transforming activity in NIH3T3 cells. EMBO J 14:5048–5059

18. Matsushima-Nishiwaki R, Okuno M, Adachi S, Sano T, Akita K, Moriwaki H, Friedman SL, Kojima S (2001) Phosphorylation of retinoid X receptor α at serine 260 impairs its metabolism and function in human hepatocellular carcinoma. Cancer Res 61 (in press)

19. Huang ME, Yu-Chen Y, Shu-Rong C, Lu MX, Zhao L, Gu LJ, Wang ZY (1988) Use of all-trans retinoic acid in the treatment of acute promyelocytic leukemia. Blood 72: 567–572

20. Yamada Y, Shidoji Y, Fukutomi Y, Ishikawa T, Kaneko T, Nakagama H, Imawari M, Moriwaki H, Muto Y (1994) Positive and negative regulations of albumin gene expression by retinoids in human hepatoma cell lines. Mol Carcinogen 10:151–158

21. Nagy L, Thomazy VA, Heyman RA, Davies PJA (1998) Retinoid-induced apoptosis in normal and neoplastic tissues. Cell Death Differ 5:11–19

22. Hong WK, Lippman SM, Hittelman WN, Lotan R (1995) Retinoid chemoprevention of aerodigestive cancer: from basic research to the clinic. Clin Cancer Res 1:677–686

23. Lotan R (1996) Retinoids in cancer in cancer chemoprevention. FASEB J 10:1031–1039

24. Lotan R (1995) Retinoids and apoptosis: implications for cancer chemoprevention and therapy. J Natl Cancer Inst 22:1655–1657

Signal Transduction Involved in Cell Cycle Regulation of Normal Hepatocyte and Hepatoma Cells

Yutaka Sasaki[1], Masayoshi Horimoto[2], and Norio Hayashi[1]

Summary. The cell cycle is controlled by cyclin/cyclin-dependent kinase (CDK) complexes and CDK inhibitors. The expression and function of these molecules are induced and modulated by signal transduction governing cell growth and differentiation. We found that when normal rat hepatocyte growth was induced by hepatocyte growth factor (HGF), HGF receptor (MET) was activated, with association of MET with growth factor receptor-bound protein (GRB)-2, which is the initial step of activation of the mitogen-activated protein kinase (MAPK) cascade. We also report that the MAPK cascade was sequentially activated, followed by hepatocyte growth. Before cell density reached confluence, p21/WAF1, a CDK inhibitor, prevented Cdk4 activity, resulting in cell cycle arrest. Thus, in normal hepatocyte growth, the cell cycle is strictly controlled. In contrast, in human hepatocellular carcinomas (HCCs), MAPK/ERK (extracellular signal regulated kinase) was constitutively upregulated. Enhanced association of GRB-2 with growth factor receptors or Sos, a *ras* activator, was also observed, which may account for MAPK activation. On the other hand, p21/WAF1 in human HCCs was deficient or overwhelmed by CDKs in excess stoichiometry. These observations indicate that constitutively active MAPK and/or disturbance of cell cycle regulators contributes much to unrestricted cell growth in human HCCs. Finally, we introduced deletion mutants of GRB-2 into HepG2 cells, resulting in reduction of MAPK/ERK activity and prevention of cell growth. In addition, when we introduced the p21/WAF1 expression vector into HepG2 cells, the cell cycle was arrested at G_1/S transition with inhibition of cell growth. These results may provide a new strategy for anticancer therapy by modulating cell cycle machinery.

Key words. Signal transduction, MAPK, Cell cycle, HCC, p21/WAF1

Introduction

The majority of the cells in an adult are considered to be in a resting, or quiescent, state. When suitable extracellular cues are present, for example, during a response to injury or surgical resection, cells leave the quiescent state (G_0 phase) and enter the G_1

---

[1] Molecular Therapeutics and [2] Internal Medicine and Therapeutics, Osaka University Graduate School of Medicine, 2-2 Yamadaoka, Suita, Osaka 565-0871, Japan

phase of the cell cycle. Extracellular cues that mediate cell cycle transition can be divided into two groups: mitogenic growth factors and extracellular matrix (ECM). Mitogenic growth factors act mainly in the G_1 phase to mediate cell cycle progression over the restriction point, in terms of inducing cyclins or enhancing association of cyclins and CDKs (cyclin-dependent kinases) [1]. In contrast, CDK inhibitors and INK4 families interact with CDK and inhibit kinase activity, resulting in cell cycle arrest.

On the other hand, the extracellular membrane (ECM) exerts profound control over cells [2]. Effects of the matrix are primarily mediated by integrins, a family of cell-surface receptors which attaches cells to the matrix and mediates mechanical and chemical signals from the matrix. These signals regulate the activities of growth factor receptors and cytoplasmic kinases including mitogen-activated protein kinase (MAPK) and jun kinase (JNK), leading to induction of cyclin and suppression of CDK inhibitors. In this way, mitogenic growth factors and the ECM cooperate to regulate cell cycle progression.

Regulation of the cell cycle plays a crucial role in normal cell growth and differentiation and also in tumor progression. In this chapter, we focus on how intracellular signal transduction participates in cell cycle regulation of normal hepatocyte growth as well as liver cancer cells in vivo and in situ.

Activation of MAPK/ERK and Expression of Cell Cycle-Related Proteins in Hepatocyte Growth

Growth factors can bind to their specific receptors. Once the receptor or its substrate is phosphorylated (activated), growth factor receptor-bound protein 2 (GRB-2) binds to an activated receptor such as hepatocyte growth factor receptor (MET), epidermal growth factor receptor (EGFR), or insulin receptor substrate 1 (IRS-1), which is phosphorylated by insulin receptor β-subunit kinase [3,4]. GRB-2 was originally identified as a 25-kDa adaptor molecule containing one Src homology 2 (SH2) domain and two flanking SH3 domains, and it binds to phosphotyrosine residues found in activated growth factor receptors or their substrates. Subsequently, GRB-2 activates Ras by forming a complex with mammalian son of sevenless (Sos), a *ras* activator, leading to activation of the MAPK cascade [5] (Fig. 1a). In this way, GRB-2 binding to an activated growth factor receptor or its substrate is the initial event of MAPK cascade activation.

As a normal cell growth model, the effect of hepatocyte growth factor (HGF) on MAPK cascade activation was examined in primary cultured rat hepatocytes [6]. Tyrosyl phosphorylation (activation) of MET was enhanced 2.5 fold over the initial level, peaking at 5 min after exposure of the hepatocytes to HGF (10 ng/ml). Association of MET with GRB-2 and association of GRB-2 with Sos increased 1.6- and 1.5 fold, respectively. Increased association was observed almost in parallel with enhancement of tyrosyl phosphorylation of MET, which indicated that the HGF growth signal was transduced from activated MET through MAPK cascade. In this regard, Raf-1 and MAPK were sequentially activated after HGF stimulation; Raf-1 activity peaked at 10 min, followed by MAPK/ERK-1 activation with a peak at 20 min after stimulation (Fig. 1b).

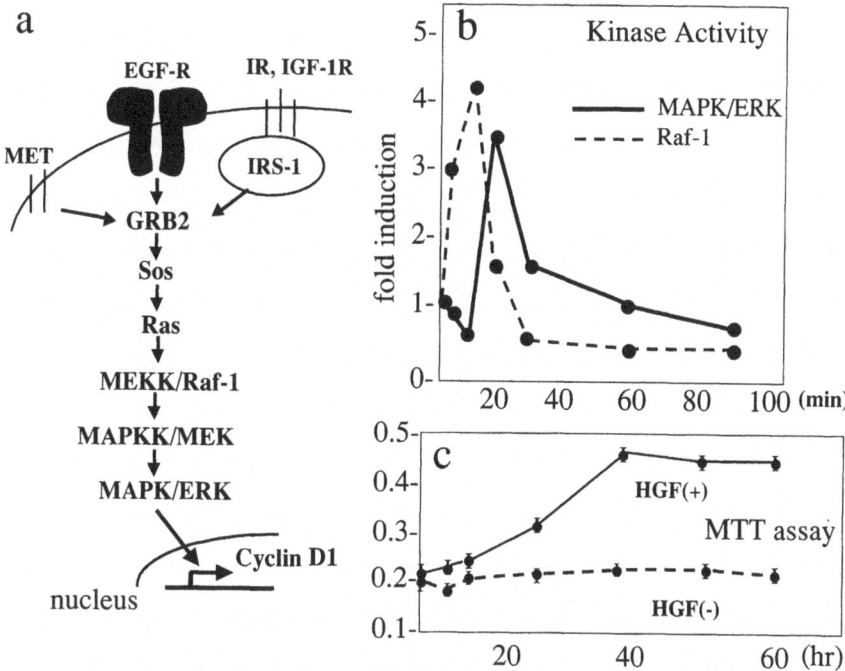

Fig. 1a–c. Activation of mitogen-activated protein kinase (MAPK)/extracellular signal regulated kinase (ERK) cascade induced by hepatocyte growth factor (HGF) stimulation. a MAPK cascade. *MET*, hepatocyte growth factor (HGF) receptor; *IRS-1*, insulin receptor substrate 1; *EGFR*, epidermal growth factor receptor; *IR*, insulin receptor; *IGF-1R*, insulin-like growth factor 1 receptor; *GRB-2*, growth factor receptor-bound protein 2; *Sos*, son of sevenless gene. b Raf-1 (*dashed line*) and MAPK/ERK (*solid line*) activation induced by HGF stimulation (10 ng/ml). Kinase activities are expressed as fold induction over initial level. Similar results were obtained in three independent experiments. c Rat hepatocyte growth induced by HGF stimulation (10 ng/ml) (*solid line*). Cell growth was assessed by MTT (*dashed line*) assay. Results were expressed as mean ± SD of 12 wells

Finally, activated MAPK/ERK-1 is translocated to the nuleus, where it induces transcriptional factors, including c-Fos and c-Jun. In turn, these two transcriptional factors consist of AP-1, and bind to AP-1-binding sites of promoter regions to induce transcription of the genes, including cyclin D_1 [7–9], which is required for cell cycle progression in the G_0/G_1 phase as a G_1 cyclin (see Fig. 1a). MAPK/ERK-1 also enhances the binding affinity of CDK with cyclin D_1. Once the cyclin D_1 and CDK complex is formed, it can phosphorylate retinoblastoma protein (RB), one of the suppressor gene products. Consequently, a transcriptional factor, E2F, is released from phosphorylated RB and restores its function. These E2F families have been implicated in the regulation of growth-promoting genes required for DNA synthesis [10–13]; E2F induces gene expression involved in the S_1 phase (Fig. 2).

Our study has shown that MAPK/ERK activation induced by HGF was accompanied by hepatocyte growth, and eventually the cell growth curve reached a plateau about 38 h after HGF stimulation (Fig. 1c), indicating that the cell cycle may be strictly

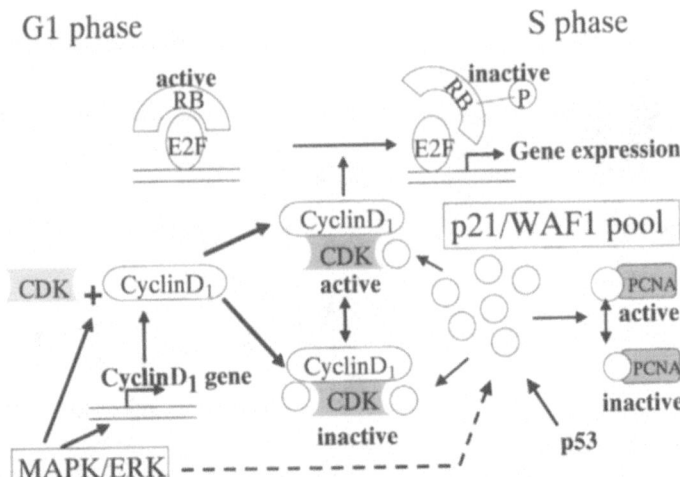

Fig. 2. Roles of cyclin/cyclin-dependent kinase (*CDK*) complex and CDK inhibitor in the cell cycle. MAPK/ERK enhances the binding affinity of CDK with cyclin D_1. Once the cyclin D_1 and CDK complex is formed, it can phosphorylate retmoblastoma protein (*RB*). Consequently, a transcriptional factor E2F is relieved from phosphorylated RB and induces gene expression involved in S_1 phase. On the other hand, p21/WAF1 mainly functions as a G_1-CDK inhibitor, and the ratio of p21/WAF1 to CDK is critical for the function as a CDK inhibitor. Also, p21/WAF1 can bind to proliferating cell nuclear antigen (*PCNA*) and block its ability to activate DNA polymerase δ, affecting DNA replication. In addition to its induction by p53, p21/WAF1 is upregulated on entry into the cell cycle after stimulation with mitogens

regulated after HGF stimulation. Among a variety of cell cycle regulators, we have examined p21/WAF1, which is one of the CDK inhibitors, because it has been shown that, in addition to its induction by antimitogenic agents, including transforming growth factor (TGF)-β, p21/WAF1 is upregulated on entry into the cell cycle after stimulation with mitogens [14]. Furthermore, G_1 phase induction of p21/WAF1 is dependent on activation of MAPK/ERK [15]. On the other hand, p21/WAF1 can bind to proliferating cell nuclear antigen (PCNA) and, when present in excess stoichiometry, can block its ability to activate DNA polymerase δ, affecting DNA replication through this interaction [16] (Fig. 2).

In our study, we determined how p21/WAF1 might act on cell cycle regulation during hepatocyte growth. The association of p21/WAF1 with Cdk4 or with PCNA was not detected from 6h to 24h, to 36h, respectively, after stimulation, whereas expression of cyclin D_1 was enhanced during almost the same time period (Fig 3). In other words, when the stimulated hepatocytes were proliferating, cyclin D_1 promoted cell growth and p21/WAF1 did not bind to the Cdk4/cyclin D_1 complex to prevent Cdk activity. Afterward, p21/WAF1 associated with Cdk4 and restored its inhibitory action before the cell density reached confluence. A previous study found that p21/WAF1 mRNA expression increased with a peak at 5h after HGF stimulation [14]. Our observations, together with the previous work, indicate that p21/WAF1 protein may start to take effect while cells are proliferating.

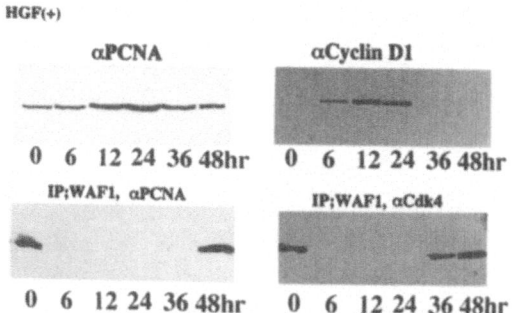

Fig. 3. Association of p21/WAF1 with cell cycle regulators in hepatocyte growth induced by HGF stimulation. PCNA expression or cyclin D_1 expression was examined by Western blotting with its specific antibody (*upper left panel* and *upper right panel*, respectively). Association of p21/WAF1 with PCNA or with Cdk4 is shown in the *lower left panel* and *lower right panel*, respectively. *Numbers below* the panels indicate hours after HGF stimulation (10 ng/ml)

The possible involvement of increased p21/WAF1 expression in the immediate to early G_1 phase has been proposed; p21/WAF1 works by a stoichiometric mechanism in which low levels may activate CDK activity and high levels inhibit CDK activity [15–17]. Therefore, p21/WAF1 may, at a low concentration, promote the cell cycle, and alternatively it may prevent the transition of premature cells through G_1 phase at a high concentration.

In any case, the MAPK cascade plays an important role in normal hepatocyte proliferation by mediating cell cycle progression in terms of modulating p21/WAF1 function.

Activation of MAPK/ERK and Induction of Cell Cycle-Related Genes in HCC and Chronic Liver Disease

It has been reported that MAPK/ERK activation is essential for oncogenic transformation [18]. Therefore, we examined MAPK/ERK activities in human HCCs; MAPK/ERK activities in HCCs were significantly higher ($P = 0.0146$, paired t test) than those in adjacent noncancerous lesions [19]. When the noncancerous lesions were classified histologically into chronic hepatitis or liver cirrhosis, MAPK/ERK activities were upregulated with the progression of liver disease. In addition, MAPK/ERK activities in HCCs were enhanced more than in chronic hepatitis or in liver cirrhosis. As a downstream event, expression of cyclin D_1 protein was upregulated with the progression of liver disease (Fig. 4). Since a previous study showed that cyclin D_1 expression was induced by DNA amplification of the cyclin D_1 gene in human HCCs [20], DNA amplification of cyclin D_1 was assessed using Southern blot analysis in 25 cases. Cyclin D_1 gene amplification in a cancerous lesion was observed in 3 of the 25 HCCs (12%), which agrees closely with the previous report. On the contrary, the remaining 22 of the 25 cases of HCC (88%) exhibited no genomic DNA amplification. In these 22 cases, a significant positive correlation was obtained not only between

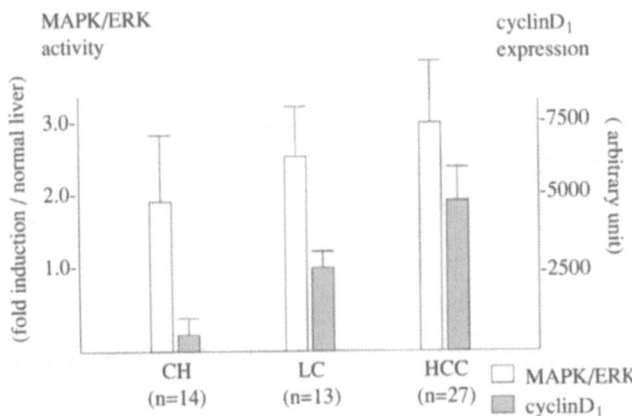

Fig. 4. MAPK/ERK activation and cyclin D$_1$ expression in chronic liver diseases. MAPK/ERK activities were expressed as fold induction over normal liver. *CH*, chronic hepatitis; *LC*, liver cirrhosis; *HCC*, hepatocellular carcinoma

MAPK/ERK activity and c-Fos expression ($r_s = 0.569$; $P = 0.00588$) but also between c-Fos expression and cyclin D$_1$ expression ($r_s = 0.574$; $P = 0.00531$) in HCCs. In addition, there was a significant positive correlation between MAPK/ERK activity and cyclin D$_1$ protein expression in HCCs ($r_s = 0.497$; $P = 0.0186$) [19]. These results indicate that MAPK/ERK activation in HCCs may lead to cell proliferation, in part through overexpression of cyclin D$_1$.

Relationship Between Association of GRB-2 with Signal Molecules and MAPK/ERK Activation in Human HCCs

To clarify the mechanisms whereby MAPK/ERK was activated in human HCCs, we determined the association of GRB-2 protein with growth factor receptors. Tumor to nontumor (T/N) ratios of MAPK/ERK activity were significantly correlated with those of IRS-1/GRB-2 association ($P < 0.01$) or that of the GRB-2/Sos association ($P < 0.05$) (data not shown). However, there was no significant correlation of T/N ratios of MAPK/ERK activities with those of MET/GRB-2 association or EGFR/GRB-2 association. These observations led us to speculate that functional modulation of GRB-2-mediated signaling may prevent cancer cell growth through regulation of MAPK/ERK activity.

Cell Cycle Machinery of Human HCCs

There have been a variety of studies indicating that the cell cycle is dysregulated owing to disturbance of the cell cycle machinery in cancer cells. With regard to human HCCs, deletion or mutation of the p53 gene as well as the Rb gene has been reported [21]. On the other hand, p21/Waf1 is induced in not only p53-dependent but also p53-independent fashion [22,23]. Mutation of p21/Waf-1 was barely detected in a variety of malignant tumors [24,25], but decrease or deletion of p21/WAF1 protein expression was reported in tumors including HCCs [26,27].

In this context, we examined the expression of cell cycle regulators in six representative cases of human HCCs. There was no apparent difference in p53 protein expression between NT and T. p21/WAF1 expression was barely detected in cases 1, 5, and 6, but the other cases exhibited p21/WAF1 in NT as well as in T, as shown in cases 2, 3, and 4. There seems to be a tendency in those cases for p21/WAF1 expression to be higher in T than in NT. Furthermore, in such cases, expression of CDK2 and/or CDK4 was higher in T than in NT as well. As discussed, p21/WAF1 functions mainly as a CDK inhibitor, and the ratio of p21/WAF1 to CDK is critical for function as a CDK inhibitor [28]. Taken together, our observations indicate that although p21/WAF1 is expressed in tumorous lesions, CDK2 or CDK4 may be overwhelming p21/WAF1 function. Regardless of the mechanisms that induce p21/WAF1 expression, reduced expression may be involved in carcinogenesis and tumor growth of HCCs.

Effect of GRB-2 Deletion Mutants or p21/WAF1 Overexpression on Cell Growth of HepG2 Cells

On the basis of the foregoing observations, we attempted to regulate cell growth of cancer cells by regulating the MAPK cascade or cell cycle machinery. First, we constructed two types of deletion mutants of GRB-2, mutant A and mutant B, in which the amino-terminal SH3 domain or the carboxyl-terminal SH3 domain, respectively, was deleted. Then, we tested whether these deletion mutants could interfere with binding of the endogenous GRB-2 to signaling molecules in human hepatoma cell line, HepG2 cells. When each deletion mutant was transfected, amount of the endogenous GRB-2 that coprecipitated with MET decreased, compared with that in the cells transfected with mock vector. Consistently, in HepG2 cells transfected with each mutant, endogenous GRB-2 coprecipitated with EGFR or IRS-1 decreased, compared with that in the cells transfected with mock control vector. Tyrosyl phosphorylation of MET, EGFR, and IRS-1 was comparable in the cells transfected with mock control vector and each deletion mutant of GRB-2.

When activity of MAPK/ERK was analyzed in the cells transfected with each deletion mutant, MAPK/ERK activity was dramatically reduced in cells transfected with mutant A or mutant B, compared with that in cells transfected with a mock control vector (Fig. 5a). These results indicate that interference of the interaction between MET, EGFR, or IRS-1 and endogenous GRB-2 results in downregulation of MAPK/ERK activity. The growth rate of the cells transfected with each deletion mutant was significantly lower than those transfected with the mock control vector (Fig. 5b). In this way, deletion mutants of GRB-2 can regulate MAPK/ERK activity, resulting in prevention of growth of cancer cells.

In the next study, we attempted to control the cell cycle of HepG2 cells, which express small amount of p21/WAF1 protein [29], by introducing p21/WAF1 expression vector. We transfected p21/WAF1 expression vector into HepG2 cells on day 1 and determined cell growth. Growth suppression was observed until day 4 after the

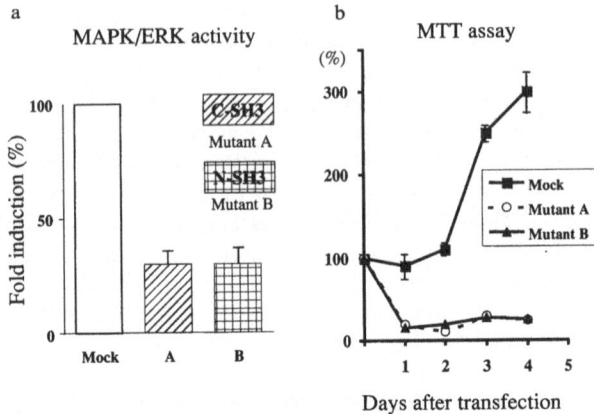

Fig. 5a,b. Effect of deletion mutants of GRB-2 on MAPK/ERK activity and cell growth of HepG2 cells. **a** HepG2 cells were homogenated 2 days after the transfection, and 400 µg of cell lysates were subjected to MAPK assay. Data are expressed as mean ± SD for three separate experiments. **b** Cell growth was determined after the transfection with deletion mutants of GRB-2 or mock control vector. The number of cells was determined by MTT assay on the indicated days after transfection. *Vertical axes* indicate arbitrary unit of the absorbance reader. Values are expressed as mean ± SD of six different wells. Similar results were obtained in three separate experiments

transfection, compared to the cells transfected with mock control vector (Fig. 6). However, cell growth was restored from day 5, owing to the degradation of p21/WAF1 protein.

Although the cell cycle profile of the parental cells displayed asynchronicity on day 0, cells after transfection displayed G_1 arrest; overexpression of p21/WAF1 increased the G_0/G_1 phase cell population to 60% 3 days after transfection (day 4), with S or G_2/M phase being 18% and 22%, respectively (Fig. 6). These findings, together with the evidence that anticancer agents work in a cell cycle-dependent fashion, suggest that combination of cell cycle synchronization with anticancer agents may enhance the effect of anticancer agents.

The final attempt was conducted to transfect p21/WAF1 expression vector transiently in HepG2, followed by addition of mitomycin C (MMC). MMC can act at the G_1/S transition and inhibits DNA synthesis by cross-linking DNA at the guanine and adenine residues [30]. As described, synchronized HepG2 cells ended G_1 arrest on day 5 after the transfection. When we transfected p21/WAF1 expression vector on day 1 and added MMC to the medium on day 4, cell growth was strikingly inhibited compared with that of cells without MMC. In contrast, cell growth was slightly inhibited by MMC compared with the mock control cells (Fig. 7). These findings indicate that anticancer agents can exhibit their actions effectively on the cells that progress in the cell cycle after synchronization by p21/WAF1 expression.

HepG2

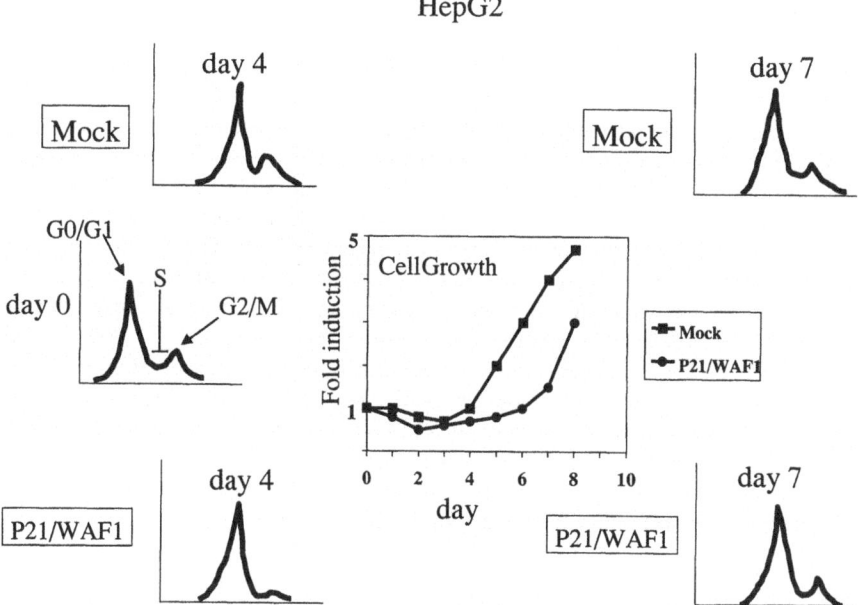

Fig. 6. Cell growth and cell cycle profile in the HepG2 cells transfected with p21/WAF1 expression vector. One hundred thousand HepG2 cells were seeded on plastic 96-well dishes in Dulbecco's modified Eagle's essential medium (DMEM) containing 10% fetal calf serum. p21/WAF1 expression vector was transfected on day 1, and cell numbers were counted in the MTT assay. Cell cycle profile was determined on day 0, day 4, and day 7. *Vertical axis* indicates fold induction of cell number over initial level. Data are expressed as mean of 12 wells of plastic 96-well dish

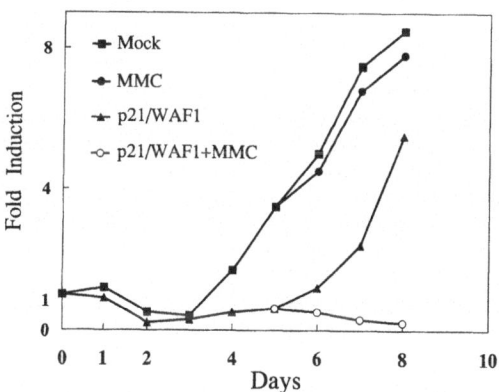

Fig. 7. Effect of combination of p21/WAF1 expression and the anticancer agent mitomycin C (MMC), on cell growth of HepG2 cells. One hundred thousand HepG2 cells were seeded on plastic 96-well dishes in DMEM containing 10% fetal calf serum, and counted with MTT assay under the indicated conditions. *Vertical axis* indicates fold induction of cell number over initial level. Data are expressed as mean of 12 wells. *MMC*, cells treated with MMC from day 4 (•); p21/WAF1, cells transfected with p21/WAF1 expression vector on day 1(▲); p21/WAF1 + MMC, cells transfected with p21/WAF1 expression vector on day 1 and treated with MMC from day 4 (○)

Conclusion

In normal hepatocyte growth, the MAPK cascade is activated in response to mitogen stimulation, associated with transient expression of cylin D_1. Consequently, hepatocytes continue proliferating until a CDK inhibitor is induced to prevent the activity of the cyclin/CDK complex. In this way, MAPK and CDK inhibitors cooperate so that the cell cycle is strictly regulated. In contrast, in human HCCs the MAPK cascade is constitutively upregulated, accompanied by cyclin D_1 overexpression. In addition, the CDK inhibitor is deficient or functionally disturbed. As these intracellular events are combined, the cell cycle accelerates out of regulation, leading to unrestricted cell growth in liver cancer cells. On the basis of these observations, reduction of MAPK or a supplement of p21/WAF1 restore cell cycle arrest and suppress the growth of liver cancer cells. In addition, our attempt to combine overexpression of p21/WAF1 with an anticancer agent dramatically enhanced its efficiency.

Further studies are being conducted to develop a new strategy for anticancer therapy in terms of modulating the cell cycle machinery.

References

1. Hunter T, Pines J (1994) Cyclins and cancer II: cyclin D_1 and CDK inhibitors come of age. Cell 79:573–582
2. Assoian RK (1993) Anchorage-dependent cell cycle progression. J Cell Biol 136: 1–4
3. Lowenstein EJ, Daly RJ, Batzer AG, et al (1992) The SH2 and SH3 domain-containing protein GRB-2 links receptor tyrosine kinases to ras signaling. Cell 70:431–442
4. Egan SE, Giddings BW, Brooks MW, et al (1993) Association of Sos Ras exchange protein with Grb2 is implicated in tyrosine kinase signal transduction and transformation. Nature 363:45–51
5. Skolnik EY, Batzer A, Li N, et al (1993) The function of GRB2 in linking the insulin receptor to Ras signaling pathways. Science 260:1953–1955
6. Wada S, Sasaki Y, Horimoto M, et al (1998) Involvement of growth factor receptor-bound protein-2 in rat hepatocyte growth. J Gastroenterol Hepatol 13:635–642
7. Chen R-H, Sarnecki C, Blenis J (1992) Nuclear localization and regulation of erk- and rsk-encoded protein kinases. Mol Cell Biol 12:915–927
8. Seth A, Gonzalez FA, Gupta S, et al (1992) Signal transduction within the nucleus by mitogen-activated protein kinase. J Biol Chem 267:24796–24804
9. Herber B, Truss M, Beato M, et al (1994) Inducible regulatory elements in the human cyclin D_1 promoter. Oncogene 9:1295–1304
10. Chellappan SP, Hiebert S, Mudryj M, et al (1991) The E2F transcription factor is a cellular target for the RB protein. Cell 65:1053–1061
11. Hielbert SW, Chellappan SP, Horowitz JM, et al (1992) The interaction of RB with E2F coincides with an inhibition of the transcriptional activity of E2F. Genes Dev 6:177–185
12. Nevins JR (1992) E2F: a link between the Rb tumor suppressor protein and viral oncoproteins. Science 258:424–429
13. Lees JA, Saito M, Vidal M, et al (1993) The retinoblastoma protein binds to a family of E2F transcription factors. Mol Cell Biol 13:7813–7825
14. Albrecht JH, Meyer AH, Hu MY (1997) Regulation of cyclin D_1-dependent kinase inhibitor p21/WAF1/Cip1/Sdi1 gene expression in hepatic regeneration. Hepatology 25:557–563

15. Gartel AL, Serfas MS, Tyner AL (1996) P21-negative regulator of the cell cycle. Proc Soc Exp Biol Med 213:138–149
16. Zhang H, Hannnon GJ, Beach D (1994) p21-containing cyclin kinases exist in both active and inactive states. Genes Dev 8:1750–1758
17. Nakanishi M, Adami GR, Robertoye RS, et al (1995) Exit from G_0 and entry into the cell cycle of cells expressing p21/Sid1 antisense RNA. Proc Natl Acad Sci USA 92:4352–4356
18. Cowley S, Paterson H, Kemp P, et al (1994) Activation of MAP kinase is necessary and sufficient for PC12 differentiation and for transformation of NIH 3T3 cells. Cell 77:841–852
19. Ito Y, Sasaki Y, Horimoto M, et al (1998) Activation of mitogen-activated protein kinases/extracellular signal regulated kinases in human hepatocellular carcinoma. Hepatology 27:951–958
20. Nishida N, Fukuda Y, Komeda T, et al (1994) Amplification and overexpression of the cyclin D1 gene in aggressive human hepatocellular carcinoma. Cancer Res 54:3107–3110
21. Bressac B, Kew M, Wands J, et al (1991) Selective G to T mutation of p53 gene in hepatocellular carcinoma from southern Africa. Nature 350:429–431
22. Xiong Y, Hannon GJ, Zhang H, et al (1993) p21 is a universal inhibitor of cyclin kinases. Nature 366:701–704
23. Michiell P, Chedid M, Lin D, et al (1994) Induction of WAF1/CIP1 by a p53-independent pathway. Cancer Res 54:3391–3395
24. Shiohara M, El-Deiry WS, Wada M, et al (1994) Absence of WAF1 mutations in a variety of human malignancies. Blood 11:3781–3784
25. Friedman SL, Shaulian E, Littlewood T, et al (1997) Resistance to p53 mediated growth arrest and apoptosis in Hep3B hepatoma cells. Oncogene 15:63–70
26. Hui A-M, Kanai Y, Sakamoto M, et al (1997) Reduced p21WAF1/CIP1 expression and p53 mutation in hepatocellular carcinomas. Hepatology 25:575–579
27. Sasaki K, Sato T, Kurose A, et al (1997) Immunohistochemical deletion of p21WAF1/Cip1/sdi1 and p53 proteins in formalin-fixed, paraffin-embedded tissue sections of colorectal carcinomas. Hum Pathol 27:912–916
28. Harper JW, Adami GR, Wei N, et al (1993) The p21 CDK-interacting protein Cip1 is a potent inhibitor of G_1 cyclin dependent kinases. Cell 75:805–816
29. Puisieux A, Galvin K, Troalen F, et al (1993) Retinoblastoma and p53 tumor suppressor genes in human hepatoma cell lines. FASEB J 7:1407–1413
30. Knudsen KE, Booth D, Naderi S, et al (2000) RB-dependent S-phase response to DNA damage. Mol Cell Biol 20:7751–7763

Effects of Simultaneous Treatment with Growth Factors on DNA Synthesis, MAPK Activity, and G_1 Cyclins in Rat Hepatocytes

Akio Ido, Akihiro Moriuchi, Shuichi Hirono, and Hirohito Tsubouchi

Summary. Several growth factors act in concert during liver regeneration in vivo. Once hepatic injury occurs, liver regeneration is stimulated by hepatocyte growth factor (HGF), transforming growth factor (TGF)-α, and heparin-binding epidermal growth factor-like growth factor (HB-EGF), whereas TGF-β_1 terminates liver regeneration. Treatment with a combination of HGF and epidermal growth factor (EGF), in comparison with that with either HGF or EGF, was found to additively stimulate mitogen-activated protein kinase (MAPK) activity and cyclin D_1 expression additively, resulting in additive stimulation of DNA synthesis. In contrast, TGF-β_1 treatment completely inhibited DNA synthesis through a marked decrease in cyclin E expression, although growth factor-induced MAPK activity and cyclin D_1 expression were not affected by TGF-β_1. These results indicate that potent mitogens such as HGF, TGF-α, and HB-EGF can cooperatively induce the additive enhancement of liver regeneration through an increase in Ras/MAPK activity followed by cyclin D_1 expression, and that TGF-β_1 suppresses the growth factor-induced signals between cyclin D_1 and cyclin E, resulting in inhibition of DNA synthesis.

Key words. Hepatocyte growth factor, Epidermal growth factor, Transforming growth factor-β, Liver regeneration, Signal transduction

Introduction

In adult animals, hepatocytes are highly differentiated and rarely divide under normal conditions. However, under certain physiopathological stress situations, such as partial hepatectomy, viral infection, or toxic injury, hepatocytes are able to divide in response to loss of liver mass [1,2]. Immediately after a partial hepatectomy, hepatocytes enter into a state of prereplicative competence before they fully respond to growth factors. This priming step is an initiating event characterized by a transition from the G_0 to the G_1 phase of the cell cycle and is mediated by cytokines, including tumor necrosis factor-α and interleukin-6, and in part by changes in the extracellular matrix [3–5]. However, the progression of these initiated cells through the late G_1

Department of Internal Medicine II, Miyazaki Medical College, 5200 Kihara, Kiyotake, Miyazaki 889-1692, Japan

phase to the S phase is thought to require growth factors and to involve activation of cyclin–cyclin-dependent kinase (cdk) complexes. Therefore, a growth factor-dependent restriction point is precisely localized in the mid-late G_1 phase in primary cultured rat hepatocytes [6], and it is mediated by transforming growth factor (TGF)-α, heparin binding-epidermal growth factor-like growth factor (HB-EGF), and primarily by hepatocyte growth factor (HGF).

Growth and Growth-Inhibitory Factors for Hepatocyte Proliferation

HGF is known to be one of the major agents promoting the proliferation of hepatocytes. It was originally purified from the plasma of patients with fulminant hepatic failure [7,8] and rat platelets [9]. This protein is characterized as a pleiotropic factor acting as a mitogen, motogen, and morphogen for a variety of cultured cells through binding to its receptor (c-Met) on the cell membrane [10] (Fig. 1). TGF-α [11] and HB-EGF [12] also stimulate DNA synthesis in primary cultured rat hepatocytes through binding to the EGF receptor (EGFR) [13] (Fig. 1). Both c-Met and EGFR transmit signals to the nucleus, primarily through the Ras/mitogen-activated protein kinase (MAPK) pathway, which is followed by the expression of cyclin D_1 protein in the nucleus, driving the cell cycle from G_1 to the S phase and initiating DNA synthesis [14,15]. In contrast, TGF-β_1 is a growth-inhibitory factor for hepatocytes, completely inhibiting the effects of HGF, TGF-α, and EGF on DNA synthesis in hepatocytes [16]. As is well known, TGF-β terminates liver regeneration in the late phase. Activin,

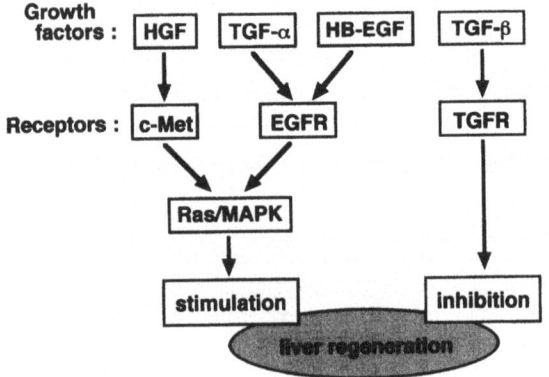

Fig. 1. Growth and growth-inhibitory factors for liver regeneration. Hepatocyte growth factor (*HGF*) promotes the proliferation of hepatocytes through binding to its receptor (*c-Met*) on the cell membrane. Transforming growth factor (*TGF*)-α [11] and heparin-binding epidermal growth factor-like growth factor (*HB-EGF*) also stimulate DNA synthesis through binding to the EGF receptor (*EGFR*). Both c-Met and EGFR transmit signals into the nucleus, primarily through the Ras/mitogen-activated protein kinase (*MAPK*) pathway. In contrast, TGF-β_1 completely inhibits the effects of HGF, TGF-α, and HB-EGF on DNA synthesis in hepatocytes through binding to its receptor (*TGFR*)

which is expressed in the normal rat liver, is another growth-inhibitory factor for hepatocytes [17]. These growth and growth-inhibitory factors act in concert during liver regeneration.

Interaction Among Growth Factors in Intracellular Signal Transduction for DNA Synthesis

In response to growth factors, Ras/MAPK, phosphatidyl inositol-3 kinase (PI3K) [18], and the Janus kinase (Jak)/signal transducer and activator of transcription (STAT) [19] are activated in hepatocytes. In these pathways, the Ras/MAPK system plays a central role and is primarily associated with cyclin D_1 expression, which is believed to play an important role during the G_1 phase in many types of cells [15,20,21].

In primary cultured rat hepatocytes, treatment with a combination of HGF and EGF has been found to induce additive stimulation of DNA synthesis through additive stimulation of MAPK activity (Fig. 2) [22]. When rat hepatocytes were labeled with proliferating cell nuclear antigen (PCNA), combined treatment with HGF and EGF increased the number of PCNA-positive cells when compared with the number noted with either HGF or EGF. These results indicate the possibility that some populations

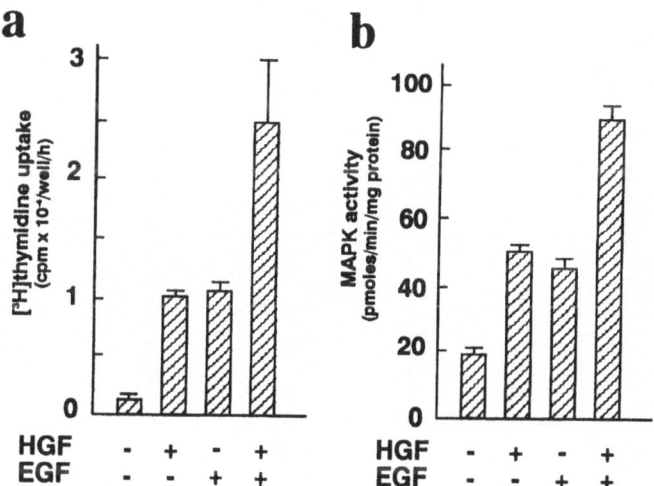

Fig. 2a,b. DNA synthesis and MAPK activity in primary cultured rat hepatocytes. **a** Effects of HGF or EGF (or both) on DNA synthesis. DNA synthesis in rat hepatocytes was measured by [³H]thymidine incorporation into DNA after incubation with HGF (10 ng/ml) or EGF (10 ng/ml) for 36 h. DNA synthesis represents the mean [³H]thymidine uptake ($n = 3$); *bars*, SE. **b** Effect of HGF or EGF on MAPK activity. Hepatocytes were treated with HGF (10 ng/ml) or EGF (10 ng/ml) for 10 min, and MAPK activity was then measured using the p42/p44 MAPK enzyme assay system (Amersham, Buckinghamshire, UK). Each *column* represents the mean p42/44 MAPK activity ($n = 3$); *bars*, SE

of hepatocytes are responsive to either HGF or EGF. Therefore, the effect of treatment with a combination of HGF and EGF would depend in part on the number of stimulated cells, resulting in the effect being additive rather than synergistic.

Where Does TGF-β_1 Inhibit the Growth Factor-Induced Signal Transductions in Primary Cultured Rat Hepatocytes?

TGF-β_1 is well known as an inhibitory growth factor in hepatocytes that mediates the arrest of the G_1 phase and is also induced in regenerating rat liver beginning 24 h after a partial hepatectomy [16]. In primary cultured rat hepatocytes, although TGF-β_1 treatment completely inhibits the DNA synthesis induced by HGF, EGF, and a combination of both, growth factor-induced MAPK activity is not affected by TGF-β_1 treatment (Fig. 3) [22].

Growth factor-induced signals for cell proliferation ultimately cause the expression of G_1 cyclins, and previous studies have suggested that cyclin D_1 is involved in the regulation of hepatocyte proliferation [21]. The expression of cyclin D_1 mRNA in rat hepatocytes is additively stimulated by combined treatment with HGF and EGF and

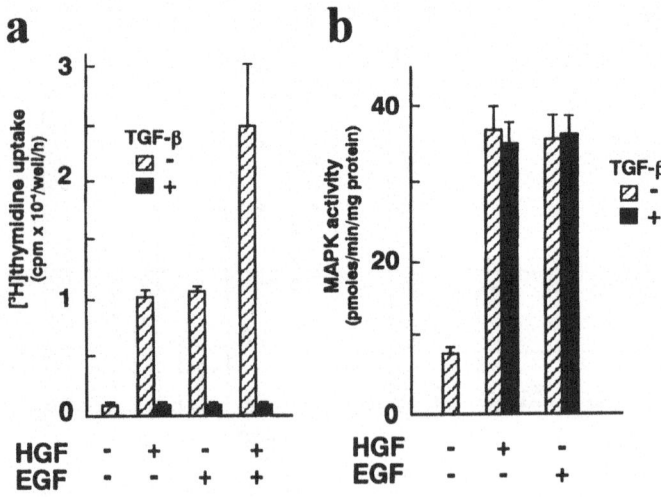

Fig. 3a,b. Effects of TGF-β_1 on DNA synthesis and MAPK activity in primary cultured rat hepatocytes stimulated by HGF or EGF (or both). **a** Effect of TGF-β_1 on HGF- or EGF-induced DNA synthesis in rat hepatocytes. Hepatocytes were plated in the absence or presence of HGF (10 ng/ml) or EGF (10 ng/ml) and treated simultaneously with 1 ng/ml of TGF-β_1. DNA synthesis represents the mean [^{3}H]thymidine uptake ($n = 3$); *bars*, SE. **b** Effects of TGF-β_1 on HGF- or EGF-induced MAPK activity in rat hepatocytes. Hepatocytes were plated in the absence or presence of HGF (10 ng/ml) or EGF (10 ng/ml) and were simultaneously treated with TGF-β_1 (1 ng/ml). MAPK activity was determined by the p42/p44 MAPK enzyme assay system. Each *column* represents the mean p42/44 MAPK activity ($n = 3$); *bars*, SE

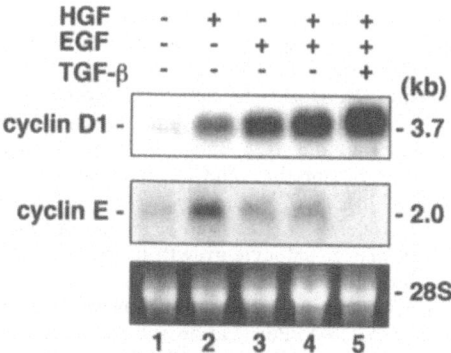

Fig. 4. Northern blot analysis of cyclin D_1 and cyclin E transcripts. Total RNA (10 μg) from unstimulated HGF- (10 ng/ml) or EGF- (10 ng/ml) stimulated hepatocytes in the absence or presence of TGF-β_1 (1 ng/ml) were analyzed using cyclin D_1 or cyclin E cDNA as a probe and 28S as a control of the total amount of RNA in each lane

was not affected by TGF-β_1 treatment (Fig. 4). In contrast, although the expression of cyclin E mRNA is also induced by HGF or EGF treatment, TGF-β_1 strongly inhibits the cyclin E expression induced by HGF or EGF (Fig. 4). These results indicate that TGF-β_1 suppresses the growth factor-induced signals between cyclin D_1 and cyclin E [22]. In some cell types, cyclin E–cdk2 activation is triggered by cyclin D_1 expression through the E2F-mediated transcription of cyclin E [23,24] or as a result of p27 (KipI), one of the major cdk inhibitors, sequestration by cyclin D_1 [25]. In our study, because cyclin E expression was strongly inhibited by TGF-β_1 treatment, TGF-β_1 could inhibit the activation of the E_2F transcription factors.

Conclusion

Potent mitogens such as HGF, TGF-α, and HB-EGF can cooperatively induce an additive enhancement of liver regeneration through an increase in Ras/MAPK activity, followed by expression of cyclin D_1 and cyclin E. In contrast, TGF-β_1 inhibits DNA synthesis by inhibiting growth factor-induced signal transductions between cyclin D and cyclin E. Thus, once hepatic injury occurs, liver regeneration is regulated by the interaction among these growth and growth-inhibitory factors.

References

1. Kennedy S, Rettinger S, Flye MW, Ponder KP (1995) Experiments in transgenic mice show that hepatocytes are the source for postnatal liver growth and do not stream. Hepatology 22:160–168
2. Rhim JA, Sandgren EP, Palmiter RD, Brinster RL (1995) Complete reconstitution of mouse liver with xenogeneic hepatocytes. Proc Natl Acad Sci USA 92:4942–4946
3. Michalopoulos GK, DeFrances MC (1997) Liver regeneration. Science 276:60–66

4. Yamada Y, Kirillova I, Peschon JJ, Fausto N (1997) Initiation of liver growth by tumor necrosis factor: deficient liver regeneration in mice lacking type I tumor necrosis factor receptor. Proc Natl Acad Sci USA 94:1441–1446
5. Cressman DE, Vintermyr OK, Doskeland SO (1996) Liver failure and defective hepatocyte regeneration in interleukin-6-deficient mice. Science 274:1379–1383
6. Loyer P, Cariou S, Glaise D, Bilodeau M, Baffet G, Guguen-Guillouzo C (1996) Growth factor dependence of progression through G1 and S phases of adult rat hepatocytes in vitro. J Biol Chem 271:11484–11492
7. Gohda E, Tsubouchi H, Nakayama H, Hirono S, Takahashi K, Koura M, Hashimoto S, Daikuhara Y (1986) Human hepatocyte growth factor in plasma from patients with fulminant hepatic failure. Exp Cell Res 166:139–150
8. Gohda E, Tsubouchi H, Nakayama H, Hirono S, Sakiyama O, Takahashi K, Miyazaki H, Hashimoto S, Daikuhara Y (1988) Purification and partial characterization of hepatocyte growth factor from plasma of a patient with fulminant hepatic failure. J Clin Invest 81:414–419
9. Nakamura T, Nawa K, Ichihara A, Kaise N, Nishino T (1987) Purification and subunit structure of hepatocyte growth factor from rat platelets. FEBS Lett 224:311–316
10. Weidner KM, Sachs M, Birchmeier W (1993) The Met receptor tyrosine kinase transduces motility, proliferation, and morphogenic signals of scatter factor/hepatocyte growth factor in epithelial cells. J Cell Biol 121:145–154
11. Mead JE, Fausto N (1989) Transforming growth factor-α may be a physiological regulator of liver regeneration by means of an autocrine mechanism. Proc Natl Acad Sci USA 86:1558–1562
12. Ito N, Kawata S, Tamura S, Kiso S, Tsushima H, Damm D, Abraham JA, Higashiyama S, Taniguchi N, Matsuzawa Y (1994) Heparin-binding EGF-like growth factor is a potent mitogen for rat hepatocytes. Biochem Biophys Res Commun 198:25–31
13. Gentry LE, Lawton A (1986) Characterization of site-specific antibodies to the erbB gene product and EGF receptor: inhibition of tyrosine kinase activity. Virology 152:421–431
14. Wada S, Sasaki Y, Horimoto M, Ito T, Ito Y, Tanaka Y, Toyama T, Kasahara A, Hayashi N, Hori M (1998) Involvement of growth factor receptor-bound protein-2 in rat hepatocyte growth. J Gastroenterol Hepatol 13:635–642
15. Talarmin H, Rescan C, Cariou S, Glaise D, Zanninelli G, Bilodeau M, Loyer P, Guguen-Guillouzo C, Baffet G (1999) The mitogen-activated protein kinase kinase/extracellular signal-regulated kinase cascade activation is a key signaling pathway involved in the regulation of G1 phase progression in proliferating hepatocytes. Mol Cell Biol 19:6003–6011
16. Braun L, Mead JE, Panzica M, Mikumo R, Bell GI, Fausto N (1988) Transforming growth factor-β mRNA increases during liver regeneration: a possible paracrine mechanism of growth regulation. Proc Natl Acad Sci USA 85:1539–1543
17. Yasuda H, Mine T, Shibata H, Eto Y, Hasegawa Y, Takeuchi T, Asano S, Kojima I (1993) Activin A: an autocrine inhibitor of initiation of DNA synthesis in rat hepatocytes. J Clin Invest 92:1491–1496
18. Royal I, Fournier TM, Park M (1997) Differential requirement of Grb2 and PI3-kinase in HGF/SF-induced cell motility and tubulogenesis. J Cell Physiol 173:196–201
19. Runge DM, Runge D, Foth H, Strom SC, Michalopoulos GK (1999) STAT 1α/1β, STAT 3 and STAT 5: expression and association with c-MET and EGF-receptor in long-term cultures of human hepatocytes. Biochem Biophys Res Commun 265:376–381
20. Sherr CJ (1994) G1 phase progression: cycling on cue. Cell 79:551–555
21. Sherr CJ (1996) Cancer cell cycles. Science 274:1672–1677
22. Moriuchi A, Hirono S, Ido A, Ochiai T, Nakama T, Uto H, Hori T, Hayashi K, Tsubouchi H (2001) Additive and inhibitory effect of simultaneous treatment with growth factors on DNA synthesis through MAPK pathway and G1 cyclins in rat hepatocytes. Biochem Biophys Res Commun 280:363–373

23. Botz J, Zerfass-Thome K, Spitkovsky D, Delius H, Vogt B, Eilers M, Hatzigeorgiou A, Jansen-Durr P (1996) Cell cycle regulation of the murine cyclin E gene depends on an E2F binding site in the promoter. Mol Cell Biol 16:3401–3409
24. Ohtani K, DeGregori J, Nevins JR (1995) Regulation of the cyclin E gene by transcription factor E2F1. Proc Natl Acad Sci USA 92:12146–12150
25. Sherr CJ, Roberts JM (1995) Inhibitors of mammalian G1 cyclin-dependent kinases. Genes Dev 9:1149–1163

Involvement of p21$^{\text{WAF1/CIP1}}$ and p27$^{\text{KIP1}}$ in Troglitazone-Induced Cell Cycle Arrest in Human Hepatoma Cell Lines

Hironori Koga and Michio Sata

Summary. Peroxisome proliferator-activated receptor-γ (PPAR-γ) regulates cell growth and differentiation. Recent evidence has suggested that PPAR-γ ligands have antitumor effects through inhibiting cell growth and inducing cell differentiation in several types of malignant neoplasm. In the present study, we investigated (a) the expression of PPAR-γ in both human hepatoma cell lines and five resected human hepatocellular carcinoma (HCC) tissues, (b) the growth inhibitory effect of troglitazone, a PPAR-γ ligand, on those hepatoma cells, and (c) the molecular mechanisms of troglitazone-induced cell cycle arrest. Five hepatoma cell lines, HLF, HuH-7, HAK-1A, HAK-1B, and HAK-5, were used. The mRNA expression levels of PPAR-γ, p21$^{\text{WAF1/CIP1}}$, and p27$^{\text{KIP1}}$ were determined by real-time quantitative reverse transcription-polymerase chain reaction. The expression of cell cycle-regulating proteins, such as p21, p27, cyclin E, and pRb, was examined using Western blotting. PPAR-γ was constitutively expressed in all the cell lines and the HCC tissues used in this study. A cytostatic effect of troglitazone was found in those cell lines, and this inhibition of cell growth was dosage dependent. G_0/G_1 arrest was demonstrated in flow cytometric analysis in HLF, HAK-1A, HAK-1B, and HAK-5, all of which showed an increased expression of p21 protein. However, HuH-7, lacking p21 protein expression, did not demonstrate clear arrest in the cell cycle analysis. HLF, which is pRb deficient, responded most profoundly to troglitazone, showing an increased expression of not only p21 but also p27. These findings suggest that p21 and p27 might be involved in troglitazone-induced cell cycle arrest in human hepatoma cells.

Key words. PPAR, Cyclin-dependent kinase inhibitor, Real-time RT-PCR, pRb

Introduction

Peroxisome proliferator-activated receptor-γ (PPAR-γ), a member of the nuclear hormone receptor superfamily, regulates growth arrest and terminal differentiation of adipocytes [1,2]. Recently, PPAR-γ ligands, some of which are clinically used as a

Second Department of Medicine and Kurume University Research Center for Innovative Cancer Therapy, Kurume University School of Medicine, 67 Asahi-machi, Kurume, Fukuoka 830-0011, Japan

new class of antidiabetic drug, have been shown to inhibit cell growth in several malignant cell types [3–5]. PPAR-γ ligands can inhibit cell growth and induce terminal differentiation of human liposarcoma cells [3,6]. PPAR-γ ligands also promote terminal differentiation and induce apoptosis of breast cancer cells [7,8]. However, such growth inhibitory effect via PPAR-γ activation has not been assessed in human hepatoma cells.

In eukaryotes, the cell cycle is tightly regulated by several protein kinases composed of a cyclin-dependent kinase (CDK) subunit(s) and corresponding regulatory cyclin subunit(s), and CDK inhibitors [9,10]. CDK inhibitors are grouped into two distinct families based on sequence homology and targets of inhibition [11]: one is the INK4 family, such as $p15^{INK4B}$, $p16^{INK4A}$, $p18^{INK4C}$, and $p19^{INK4D}$, and the other is the CIP/KIP family, including $p21^{CIP1}$, $p27^{KIP1}$, and $p57^{KIP2}$.

The protein product of the retinoblastoma tumor suppressor gene pRb is a negative regulator of cell proliferation and a potential substrate for cyclin E/CDK2 complex at the G_1- to S-phase transition of the cell cycle [12]. Hypophosphorylated pRb in G_1 is active for cell growth suppression, while its phosphorylated counterpart in $S/G_2/M$ is active for cell proliferation. Both p21 and p27 inhibit the activity of the cyclin D/CDK4, cyclin E/CDK2, and of cyclin A/CDK2 complexes, whereby the phosphorylation of pRb is blocked. In addition, p21 also blocks DNA replication depending on proliferation cell nuclear antigen, resulting in G_1 arrest.

Recently, p21, p27, and p18 have each been shown to play a crucial role in adipocyte differentiation through PPAR-γ activation [13]. In particular, p21 and p27 have been demonstrated to be involved in exit from the cell cycle into a predifferentiation state of postmitotic growth arrest during adipocyte differentiation [13]. In epithelial neoplasms, however, the involvement of these CDK inhibitors in PPAR-γ ligand-induced cell cycle arrest is largely unknown.

We demonstrate here that significantly increased expressions of p21 and p27 are involved in PPAR-γ ligand-induced G_1 arrest in human hepatoma cells, even in pRb-deficient cells.

Materials and Methods

Reagents

Troglitazone, a PPAR-γ ligand, was a kind gift from Sankyo Pharmaceuticals (Tokyo, Japan). Two other PPAR-γ ligands, ciglitazone and 15-deoxy-Δ12,14 prostaglandin J2 (15d-PGJ2), were purchased from Cayman Chemical (Ann Arbor, MI, USA) and BIOMOL Research Laboratories (Plymouth Meeting, PA, USA), respectively. Wy14643, a PPAR-α ligand, was from Cayman Chemical. Antibody against PPAR was obtained from Carbiochem-Novabiochem (San Diego, CA, USA); antibodies against p21 and p27 were from Transduction Laboratories (Lexington, KY, USA); and antibodies against cyclin E and pRb were from Pharmingen (San Diego, CA, USA). The protein assay reagents were obtained from Bio-Rad (Hercules, CA, USA).

Cell Lines and Cultures

We used five human hepatoma cell lines and one human colon cancer cell line. The hepatoma cell lines were HLF, HuH-7, HAK-1A [14], HAK-1B [14], and HAK-5 [15]. HLF and HuH-7 were obtained from the Japanese Cancer Research Resources Bank (Tokyo, Japan) and the Cancer Cell Repository, Tohoku University (Sendai, Japan), respectively. HAK-1A, HAK-1B, and HAK-5 were originally established in the Department of Pathology in our university. HCT-116, a colon cancer cell line, was used as a control because it has been demonstrated to have functional PPAR-γ [4].

Each cell line was grown in Dulbecco's modified Eagle medium (Sigma-Aldrich Japan, Tokyo, Japan) supplemented with 5% heat-inactivated (56°C, 30 min) fetal bovine serum (Gibco BRL/Life Technologies, Gaithersburg, MD, USA), 100 U/ml penicillin, 100 μg/ml streptomycin (Gibco) in a humidified atmosphere of 5% CO_2 in air at 37°C.

Growth Rate Analyses

To generate growth curves, 10^4 cells per well were seeded into 6-well culture plates and continuously cultured in the presence of 10 μM troglitazone. Cells from triplicate wells were harvested and counted on days 0, 2, 4, 6, and 8. To assess dosage responsiveness, cells were cultured in the medium containing 0 (vehicle only), 10, 20, 30, or 50 μM troglitazone for 6 days.

Cell Cycle Analysis

DNA content was assessed by staining ethanol-fixed cells with propidium iodide (PI) and monitoring by FACS Calibur (Becton Dickinson, Franklin Lakes, NJ, USA). The percentage of cells in the S, G_0/G_1, and G_2/M phases of the cell cycle was determined using ModFIT software (Verity Software House, Topsham, ME, USA).

Protein Extraction and Western Blotting

Lysates of logarithmically growing cells were centrifuged at 12 000 g for 30 min at 4°C and the supernatants were separated. Protein concentration was measured using a Bio-Rad protein assay kit. After being boiled for 5 min in the presence of 2-mercaptoethanol, samples containing cell lysate protein were subjected to Western blot analysis as previously described [16].

Real-Time Quantitative Reverse Transcription-Polymerase Chain Reaction (RT-PCR)

Total RNAs were isolated from hepatoma cell lines using the RNeasy system according to the manufacturer's instructions (Qiagen, Valencia, CA, USA). RNA quantification was performed using spectrophotometry. Real-time quantitative RT-PCR analyses for PPAR-γ mRNA and p21 mRNA were performed using an ABI PRISM 7700 Sequence Detection System instrument and software (PE Applied Biosystems, Foster City, CA, USA). This system uses the 5′-nuclease activity of *Taq* DNA polymerase to generate a real-time quantitative DNA analysis assay [17,18]. A nonextendable oligonucleotide hybridization probe with 5'-fluorescent and 3'-rhodamine (quench) moieties is present during the extension phase of the PCR. Degradation and release of the fluorescent moiety attributable to the 5'-nuclease activity result in a peak emission at 518 nm, and this peak is monitored every 7 s using a sequence detector. The increase in fluorescence is monitored during the complete amplification process (real time). β-actin, a housekeeping gene, was chosen as an internal standard to control for variability in amplification due to differences in starting messenger RNA (mRNA) concentrations. The sequences for the PCR primer pairs and fluorogenic probe (5' to 3'), respectively, that were used for each gene are as follows: β-actin, TCACCCACACTGTGCCCATCTACGA, CAGCGGAACCGCTC-ATTGCCAATGG, and FAM-ATGCCC-TAMRA-CCCCCATGCCATCCTGCGTp; PPAR-γ, ACCCAGAAAGCGATTCCTTCA, AGTGGTCTTCCATTACGGAGAGATC, and FAM-ATCGCATTCTGGCCCACCAACTTTG-TAMRA; and p21, GGACAGCAGAG-GAAGACCATGT, TGGAGTGGTAGAAATCTGTCATGC, and FAM-TGTCTTGTACC-CTTGTGCCTCGCTCA-TAMRA. The fluorogenic probes were FAM and TAMRA.

Results

Expression of PPAR-γ

We initially examined the expression of PPAR-γ in five human hepatoma cell lines using real-time quantitative RT-PCR (Fig. 1A). All five hepatoma cell lines, HLF, HuH-7, HAK-1A, HAK-1B, and HAK-5, significantly expressed PPAR-γ mRNA. To quantify the expression levels of PPAR-γ protein, we performed Western blot analysis using a specific antibody to PPAR-γ (Fig. 1B). HCT-116, a colon cancer cell line, was used as a positive control for PPAR-γ expression. All five hepatoma cell lines constitutively expressed PPAR-γ protein. However, the expression level of PPAR-γ protein did not necessarily correspond to that of PPAR-γ mRNA.

Effects of PPAR Ligands on Hepatoma Cells

Troglitazone at 10 μM inhibited cell growth in all hepatoma cell lines, with the greatest effect in HLF (Fig. 2A). HuH-7 has been demonstrated to be the poorest responder to troglitazone, although the expression levels of both PPAR-γ mRNA and protein were similar to those of HLF. This suppressive effect of troglitazone was cytostatic, but not cytotoxic, requiring several days for a clear difference in cell number. A dosage-dependent growth-inhibitory effect was found in all lines (Fig. 2B).

Cell Cycle Analysis

Cell cycle analyses of all cell lines were performed after their exposure to 50 μM troglitazone for 24 and 48 h. HLF, HAK-1A, HAK-1B, and HAK-5 showed a significant increase in the number of cells in the G_1 phase of the cell cycle (Fig. 3) but HuH-7 did not. A significant increase in the sub-G_1 cell population, representing apoptotic cells, was not demonstrated in any line using flow cytometry.

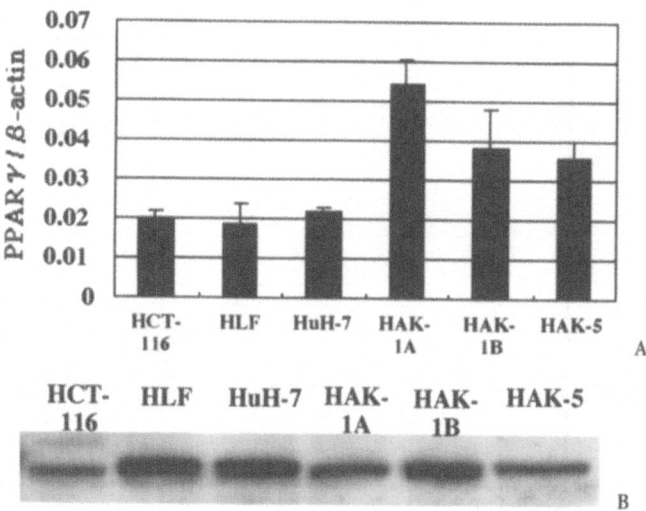

Fig. 1. Expression levels of peroxisome proliferator-activated receptor (*PPAR*)-γ mRNA (**A**) and PPAR-γ protein (**B**) in human hepatoma cell lines. PPAR-γ mRNA and protein levels were determined using real-time RT-PCR and Western blotting, respectively. All hepatoma cell lines constitutively expressed both PPAR-γ mRNA and protein, without any significant correlation between the mRNA level and the protein level in each cell line. HCT-116, a colon cancer cell line, was used as a positive control for PPAR-γ expression

Induction of p21 mRNA and Protein Expression by Troglitazone

p21 Protein expression was constitutively expressed in HAK-1A, HAK-1B, and HAK-5 but not in HLF or HuH-7. After treatment with 50 µM troglitazone for 24 h, newly induced p21 protein expression was detected in HLF, and a significantly increased expression of p21 protein was found in HAK-1A, HAK-1B, and HAK-5. HuH-7, a poor responder to troglitazone, did not express a detectable level of p21 protein even after troglitazone treatment (Fig. 4). The expression level of p21 mRNA was examined using real-time quantitative RT-PCR assay, resulting in clearly increased expression in HLF, HAK-1A, HAK-1B, and HAK-5, with a minimum increase in HuH-7 (Fig. 5).

p27 Protein Expression

An increased expression of p27 protein after 50 µM troglitazone treatment was demonstrated in HLF (Fig. 6). In the other four cell lines, the p27 protein was constitutively expressed but was not enhanced by troglitazone. HAK-1A, resembling well-differentiated HCC cells, expressed profoundly more p27 protein than HAK-1B, which is the clonally dedifferentiated sister cell line of HAK-1A.

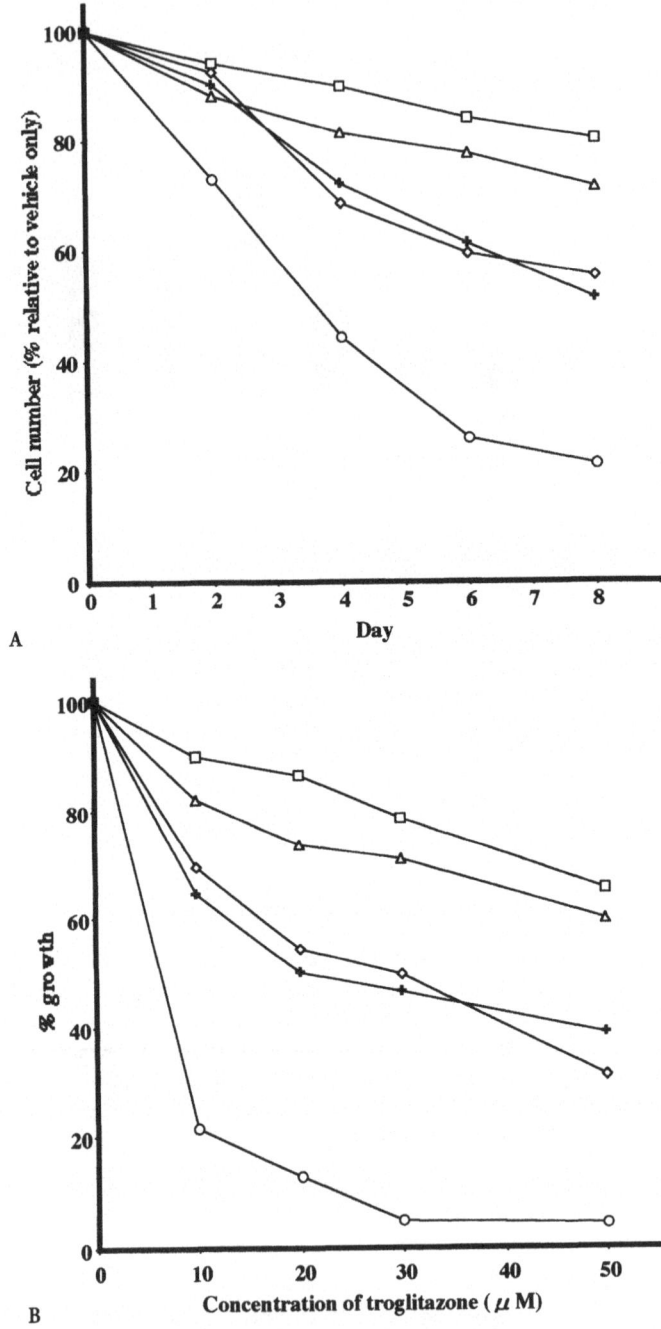

Fig. 2. A Growth curves of hepatoma cell lines continuously cultured in the presence of 10 μM troglitazone: ○, HLF; □, HuH-7; △, HAK-1A; ◇, HAK-1B; ✦, HAK-5. The average cell number from experiments in triplicate was calculated every other day until day 8, and the percentage relative to the cell number of untreated cells (vehicle only) was plotted. HLF was the most responsive cell line to troglitazone treatment, although the expression levels of PPAR-γ mRNA and protein were similar to those in HuH-7, the poorest responder to troglitazone. **B** The growth-inhibitory effect of troglitazone was demonstrated in a dosage-dependent manner, determined on day 6. ○, HLF; □, HuH-7; △, HAK-1A; ◇, HAK-1B; ✦, HAK-5. The percentage was indicated as the cell number relative to that of untreated control

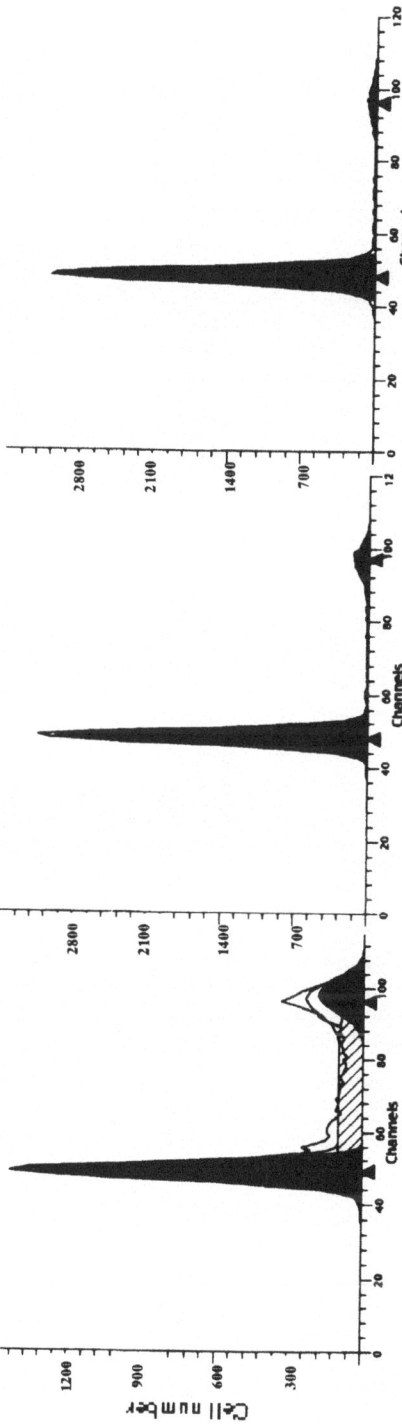

Fig. 3. Representative flow cytometric analysis for troglitazone-treated HLF cells. The cells were exposed to 50 μM troglitazone up to 48 h. Accumulation of cells in the G_1 phase of the cell cycle was found, with no increase in the sub-G_1 apoptotic cell population

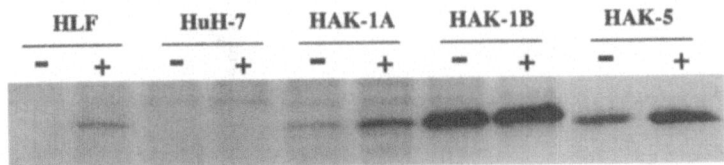

Fig. 4. Western blot analysis for p21 protein expression in hepatoma cell lines treated with 50 μM troglitazone for 24 h. Constitutive expression of p21 protein was found in HAK-1A, HAK-1B, and HAK-5, but not in HLF or HuH-7. Induction of p21 protein expression by troglitazone was demonstrated in HLF, HAK-1A, HAK-1B, and HAK-5, but not in HuH-7, which was the poorest responder to troglitazone among the five cell lines used

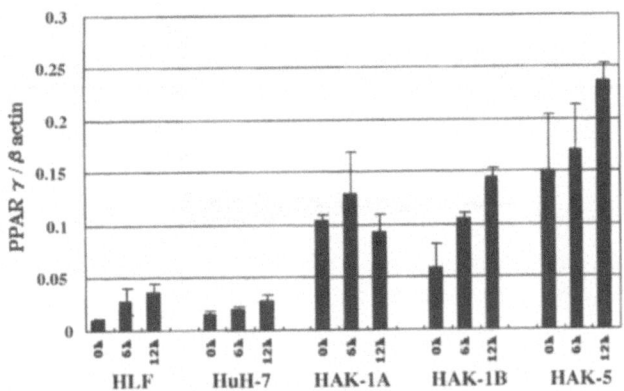

Fig. 5. The expression level of p21 mRNA was quantified using real-time RT-PCR using an ABI PRISM 7700 Sequence Detection System instrument and software. An increased expression of p21 mRNA was clearly shown in HLF, HAK-1B, and HAK-5, with a minimal increase in HuH-7. HAK-1A demonstrated a transient increase in the p21 mRNA level

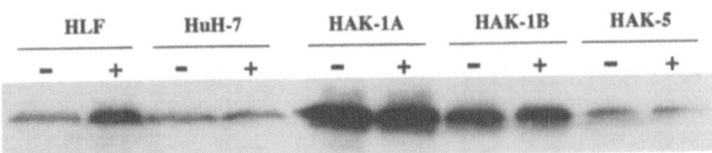

Fig. 6. Western blot analysis for p27 protein expression in hepatoma cell lines treated with 50 μM troglitazone for 24 h. An increased expression of p27 was demonstrated in HLF. No significant alteration in the level of p27 protein was shown in the other cell lines

Cyclin E Protein Expression

The expression level of cyclin E protein was examined using Western blotting, demonstrating a significantly increased expression in HLF without any significant difference in the other cell lines (Fig. 7).

pRb Expression

A significant decrease in phosphorylated pRb was demonstrated at 48 h after troglitazone treatment in HAK-1A, HAK-1B, and in HAK-5, although the decrease in HAK-1A was minimal (Fig. 8). No clear decrease in phosphorylated pRb was shown in HuH-7. HLF did not express pRb.

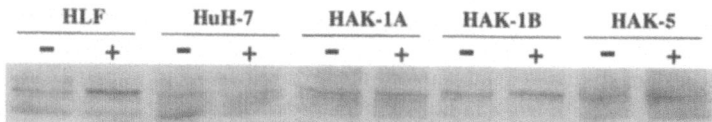

Fig. 7. The expression level of cyclin E protein was examined using Western blotting after treatment with 50 μM troglitazone for 24 h. An enhanced expression of cyclin E was detected in HLF. In the other four hepatoma cell lines, no significant difference in cyclin E protein expression was seen after troglitazone treatment

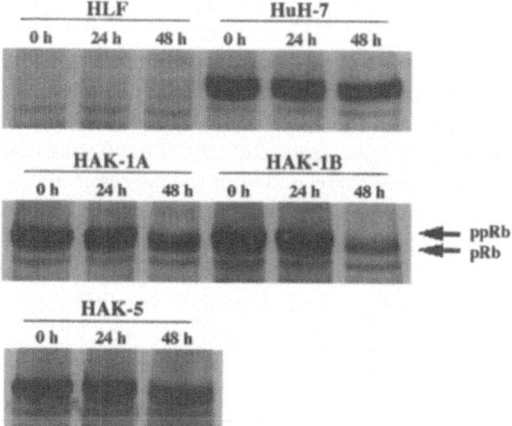

Fig. 8. Western blot analysis of pRb expression in hepatoma cell lines treated with 50 μM troglitazone for 24 h. Phosphorylated pRb (*ppRb*) was significantly decreased at 48 h after troglitazone treatment in HAK-1A, HAK-1B, and HAK-5, although the decrease in HAK-1A was minimal. No obvious decrease in phosphorylated pRb was seen in HuH-7. HLF was a pRb-deficient line

Discussion

In the present study, we first identified significant expression of both PPAR-γ mRNA and PPAR-γ protein in human hepatoma cell lines. The expression level of PPAR-γ protein for each cell line did not necessarily correspond to that of PPAR-γ mRNA or to the magnitude of the growth-inhibitory effect, suggesting the existence of its post-transcriptional posttranslational regulation in human hepatoma cell lines [19,20].

Growth curves of the cells demonstrated that HLF was the most sensitive line for troglitazone treatment and that HAK-1B, HAK-5, and HAK-1A were moderately sensitive. HuH-7 was a poor responder to troglitazone. G_1 arrest of the cell cycle was determined in HLF, HAK-1A, HAK-1B, and in HAK-5 using flow cytometric analysis. No obvious arrest of the cell cycle was evident in HuH-7 using flow cytometry. To elucidate the molecular mechanism of troglitazone-induced G_1 arrest, we initially investigated the expressions of several CDK inhibitors, such as p21 and p27, because these CDK inhibitors have been shown to be involved not only in adipocyte differentiation through PPAR-γ activation but also in epithelial growth arrest and differentiation [21].

In the present study, we demonstrated the involvement of p21 and p27 in troglitazone-induced cell cycle arrest in human hepatoma cells. Hepatoma cell lines responsive to troglitazone expressed significantly increased p21 protein, as well as p27 protein in the most responsive HLF cells. In contrast, a poor responder, HuH-7, lacked both the constitutive and the induced type of p21 protein. Thus, the growth-inhibitory effect on human hepatoma cells through PPAR-γ activation was considered to be dependent, at least in part, on the expression level of induced p21 protein but not on that of PPAR-γ. A recent report has supported this concept, demonstrating involvement of p21 in troglitazone-induced cell cycle arrest in myeloid leukemia cell lines [22]. In human HCC, reduced p21 expression has previously been demonstrated, participating in hepatocarcinogenesis [23]. From this aspect, the antitumor potential of PPAR-γ ligands, which upregulate p21 expression, should be expanded in creating promising new drugs for future cancer therapy.

We have demonstrated, for the first time, involvement of p27 in troglitazone-induced cell cycle arrest in human neoplastic cells. It has been shown that the expression level of p27 was inversely correlated with the aggressiveness of neoplasms and correlated with patient survival in several types of cancer [24–26]. In the present study, constitutive p27 protein was expressed predominantly in HAK-1A in comparison with HAK-1B, which is the aggressive and clonally dedifferentiated sister cell line of HAK-1A. Significant induction of p27 protein by troglitazone might open the way to a differentiation-induction therapy for neoplasms using PPAR-γ ligands in human HCC, although the increased expression of p27 protein has not been induced in all hepatoma cell lines used.

Cyclin E is a driving force in the G_1- to S-phase transition in the cell cycle, forming a complex with CDK2. Of interest, increased, but not decreased, expression of cyclin E protein has been paradoxically shown in HLF, which was the most sensitive cell line to troglitazone. It has been demonstrated that cyclin E was induced by genotoxic stress, such as ionizing radiation, in hematopoietic cells [27]. However, this may not be attributable solely to the possible genotoxicity of troglitazone in hepatoma cells, because an increased expression of cyclin E protein was not found in the other four hepatoma cell lines used in this study. Recent evidence has suggested that ubiquitin-dependent degradation of cyclin E was mediated by Skp2 [28]. Furthermore, Skp2 is known to be involved in the ubiquitin-dependent degradation of p27 [28]. Thus, troglitazone might contribute to accumulation not in only p27 but also in cyclin E in HLF through downregulation of Skp2.

Phosphorylated pRb is active for cell cycle progression. In this study, a significant decrease in phosphorylated pRb was demonstrated at 48 h after troglitazone treatment in HAK-1A, HAK-1B, and HAK-5, all of which expressed troglitazone-induced p21 protein, suggesting that hypophosphorylation of pRb was executed by the troglitazone-induced p21 protein through CDK2 inactivation. In contrast, no obvious decrease in phosphorylated pRb was shown in the p21 protein-deficient cell line HuH-7, resulting in poor responsiveness to troglitazone treatment. Interestingly, troglitazone induced G_0/G_1 arrest even in the pRb-deficient cell line HLF. Recent evidence has suggested the direct involvement of p21 in the inactivation of E2F transcription factors, which promote cell proliferation via E2F-responsive target genes [29]. Thus, it was speculated that troglitazone-induced p21 protein directly suppressed E2F activity in pRb-deficient HLF cells, resulting in G_1 arrest.

References

1. Tontonoz P, Hu E, Spiegelman BM (1994) Stimulation of adipogenesis in fibroblasts by PPARγ2, a lipid-activated transcription factor. Cell 79:1147–1156
2. Kliewer SA, Lenhard JM, Willson TM, Patel I, Morris DC, Lehmann JM (1995) A prostaglandin J2 metabolite binds peroxisome proliferator-activated receptor γ and promotes adipocyte differentiation. Cell 83:813–819
3. Tontonoz P, Singer S, Forman BM, Sarraf P, Fletcher JA, Fletcher CDM, Brun RP, et al (1997) Terminal differentiation of human liposarcoma cells induced by ligands for peroxisome proliferator-activated receptor γ and the retinoid X receptor. Proc Natl Acad Sci USA 94:237–241
4. Sarraf P, Mueller E, Jones D, King FJ, DeAngelo DJ, Partridge JB, Holden SA, et al (1998) Differentiation and reversal of malignant changes in colon cancer through PPARγ. Nat Med 4:1046–1052
5. Kubota T, Koshizuka K, Williamson EA, Asou H, Said JW, Holden S, Miyoshi I, et al (1998) Ligand for peroxisome proliferator-activated receptor γ (troglitazone) has potent antitumor effect against human prostate cancer both *in vitro* and *in vivo*. Cancer Res 58:3344–3352
6. Demetri GD, Fletcher CDM, Mueller E, Sarraf P, Naujoks R, Campbell N, Spiegelman BM, et al (1999) Induction of solid tumor differentiation by the peroxisome proliferator-activated receptor-γ ligand troglitazone in patients with liposarcoma. Proc Natl Acad Sci USA 96:3951–3956
7. Elstner E, Muller C, Koshizuka K, Williamson EA, Park D, Asou H, Shintaku P, et al (1998) Ligands for peroxisome proliferator-activated receptor γ and retinoic acid receptor inhibit growth and induce apoptosis of human breast cancer cells in vitro and in BNX mice. Proc Natl Acad Sci USA 95:8806–8811
8. Suh N, Wang Y, Williams CR, Risingsong R, Gilmer T, Willson TM, Sporn MB (1999) A new ligand for the peroxisome proliferator-activated receptor-γ (PPAR-γ), GW7845, inhibits rat mammary carcinogenesis. Cancer Res 59:5671–5673
9. Grana X, Reddy EP (1995) Cell cycle control in mammalian cells: role of cyclins, cyclin dependent kinases (CDKs), growth suppressor genes and cyclin-dependent kinase inhibitors (CKIs). Oncogene 11:211–219

10. Lees E (1995) Cyclin dependent kinase regulation. Curr Opin Cell Biol 7:773–780
11. Sherr CJ, Roberts JM (1995) Inhibitors of mammalian G_1 cyclin-dependent kinases. Genes Dev 9:1149–1163
12. Weinberg RA (1995) The retinoblastoma protein and cell cycle control. Cell 81:323–330
13. Morrison RF, Farmer SR (1999) Role of PPARγ in regulating a cascade expression of cyclin-dependent kinase inhibitors, p18 (INK4c) and p21 (Waf1/Cip1), during adipogenesis. J Biol Chem 274:17088–17097
14. Yano H, Iemura A, Fukuda K, Mizoguchi A, Haramaki M, Kojiro M (1993) Establishment of two distinct human hepatoma cell lines from a single nodule showing clonal dedifferentiation of cancer cells. Hepatology 18:320–327
15. Yano H, Iemura A, Haramaki M, Ogasawara S, Takayama A, Akiba J, Kojiro M (1999) Interferon alfa receptor expression and growth inhibition by interferon alfa in human liver cancer cell lines. Hepatology 29:1708–1717
16. Koga H, Sakisaka S, Ohishi M, Kawaguchi T, Taniguchi E, Sasatomi K, Harada M, et al (1999) Expression of cyclooxygenase-2 in human hepatocellular carcinoma: relevance to tumor dedifferentiation. Hepatology 29:688–696
17. Gibson UE, Heid CA, Williams PM (1996) A novel method for real time quantitative RT-PCR. Genome Res 6:995–1001
18. Fink L, Seeger W, Ermert L, Hanze J, Stahl U, Grimminger F, Kummer W, et al (1998) Real-time quantitative RT-PCR after laser-assisted cell picking. Nat Med 4:1329–1333
19. Hu E, Kim JB, Sarraf P, Spiegelman BM (1996) Inhibition of adipogenesis through MAP kinase-mediated phosphorylation of PPARγ. Science 274:2100–2103
20. Adams M, Reginato MJ, Shao D, Lazar MA, Chatterjee VK (1997) Transcriptional activation by peroxisome proliferator-activated receptor γ is inhibited by phosphorylation at a consensus mitogen-activated protein kinase site. J Biol Chem 272:5128–5132
21. Tian JQ, Quaroni A (1999) Involvement of p21 (WAF1/Cip1) and p27 (Kip1) in intestinal epithelial cell differentiation. Am J Physiol 276:C1245–C1258
22. Sugimura A, Kiriyama Y, Nochi H, Tsuchiya H, Tamoto K, Sakurada Y, Ui M, et al (1999) Troglitazone suppresses cell growth of myeloid leukemia cell lines by induction of p21WAF1/CIP1 cyclin-dependent kinase inhibitor. Biochem Biophys Res Commun 265:453–456
23. Hui AM, Kanai Y, Sakamoto M, Tsuda H, Hirohashi S (1997) Reduced p21(WAF1/CIP1) expression and p53 mutation in hepatocellular carcinomas. Hepatology 25:575–579
24. Fredersdorf S, Burns J, Milne AM, Packham G, Fallis L, Gillett CE, Royds JA, et al (1997) High level expression of p27(kip1) and cyclin D1 in some human breast cancer cells: inverse correlation between the expression of p27(kip1) and degree of malignancy in human breast and colorectal cancers. Proc Natl Acad Sci USA 94:6380–6385
25. Loda M, Cukor B, Tam SW, Lavin P, Fiorentino M, Draetta GF, Jessup JM, et al (1997) Increased proteasome-dependent degradation of the cyclin-dependent kinase inhibitor p27 in aggressive colorectal carcinomas. Nat Med 3:231–234
26. Mori M, Mimori K, Shiraishi T, Tanaka S, Ueo H, Sugimachi K, Akiyoshi T (1997) p27 expression and gastric carcinoma. Nat Med 3:593
27. Mazumder S, Gong B, Almasan A (2000) Cyclin E induction by genotoxic stress leads to apoptosis of hematopoietic cells. Oncogene 19:2828–2835
28. Nakayama K, Nagahama H, Minamishima YA, Matsumoto M, Nakamichi I, Kitagawa K, Shirane M, et al (2000) Targeted disruption of Skp2 results in accumulation of cyclin E and p27^{Kip1}, polyploidy and centrosome overduplication. EMBO J 19:2069–2081
29. Delavaine L, La Thangue NB (1999) Control of E2F activity by p21$^{WAF1/CIP1}$. Oncogene 18:5381–5392

Role of Fas Ligand in Immunopathogenesis of Hepatocellular Carcinoma

Yasunari Nakamoto[1], Shuichi Kaneko[1], Francis V. Chisari[2], Takashi Suda[3], and Kenichi Kobayashi[1]

Summary. Chronic hepatitis predisposes patients to the development of hepatocellular carcinoma (HCC), but the relative importance of the immune response in this process has not been established. The pathogenetic mechanisms potentially responsible for HCC during chronic viral hepatitis include deregulation of viral transactivators, insertional activation of cellular proto-oncogenes, and immune-mediated chronic hepatitis, perhaps enhanced by genotoxic chemicals. The current study demonstrates that chronic immune-mediated liver disease can trigger the development of HCC in hepatitis B virus transgenic mice in the absence of other procarcinogenic stimuli. Furthermore, the incidence of HCC was diminished by treatment with neutralizing antibody to Fas ligand (FasL). These results indicate that an antigen-specific, FasL-dependent immune response can initiate and sustain the process of carcinogenesis during chronic viral hepatitis.

Key words. Chronic hepatitis, Hepatocellular carcinoma, Cytotoxic T lymphocytes, Fas ligand, Transgenic mice

Introduction

Almost all cases of hepatocellular carcinoma (HCC) occur after many years of chronic hepatitis caused by a wide array of viral and nonviral etiologies in addition to hepatitis B and C viruses (HBV and HCV). Chronic active hepatitis is characterized by indolent and progressive liver cell injury with associated hepatocellular regeneration (i.e., cellular DNA synthesis) and inflammation (i.e., the production of mutagens), which could, theoretically, precipitate random genetic and chromosomal damage and lead to the development of HCC.

Transgenic mice that express the HBV large-envelope polypeptide and produce toxic quantities of subviral particles which eventually kill the hepatocyte develop a programmed response within the liver characterized by hepatocellular regenerative hyperplasia, inflammation, Kupffer cell hyperplasia, oxygen radical production, glu-

[1] First Department of Internal Medicine, Faculty of Medicine, [3] Cancer Research Institute, Kanazawa University, 13-1 Takara-machi, Kanazawa 920-8641, Japan
[2] The Scripps Research Institute, La Jolla, CA, USA

tathione depletion, oxidative DNA damage, transcriptional deregulation, and aneuploidy that inexorably progresses to HCC [1]. Collectively, those studies proved that prolonged hepatocellular injury can trigger a sequence of events that program the hepatocyte for unrestrained growth. The current study was undertaken to determine whether the same events can be triggered by a virus-specific immune response in the absence of direct viral toxicity, random integration of viral DNA, or HBV X protein expression [2].

Materials and Methods

HBV Transgenic Mice

Lineage 107-5D (official designation, Tg[Alb-1,HBV]Bri66) (inbred B10D2, H-2^d) contains the entire HBV envelope coding region (subtype ayw) under the constitutive transcriptional control of the mouse albumin promoter. These mice express the HBV small-, middle-, and large-envelope proteins in their hepatocytes; they are immunologically tolerant to hepatitis B surface antigen (HBsAg) at the T-cell level, and they display no evidence of liver disease during their lifetime although they do develop "ground glass" hepatocytes as a result of overexpression of the large-envelope protein. No X-RNA or X-protein expression is detectable in the livers of these animals (unpublished observations). Importantly, the mice develop a severe MHC class I-restricted necroinflammatory liver disease following the adoptive transfer of HBsAg-specific cytotoxic T lymphocytes (CTLs).

Thymectomy, Irradiation, Bone Marrow Reconstitution, and Spleen Cell Transfer

Male transgenic mice, 8- to 10 weeks old, were thymectomized. Seven days later, the mice were irradiated (900 cGy) from a ^{137}Cs source (Gammacell 40 irradiator; Atomic Energy of Canada, Ottawa, Canada) and then divided into two groups for bone marrow reconstitution (BMR) with nontransgenic or transgenic donor bone marrow that had been depleted of T cells by complement fixation using anti-Thy1.2 antibody (Ab) (Pharmingen, San Diego, CA, USA). Group 1 recipients, the primary experimental animals in this study, were reconstituted by the intravenous injection of 10^7 bone marrow cells collected from the femurs and tibias of syngeneic nontransgenic B10D2 (H-2^d) mice.

To control for the potential hepatocarcinogenic effect of irradiation, group 2 recipients were reconstituted with bone marrow cells derived from their immunologically tolerant inbred transgenic littermates. One week after bone marrow reconstitution, the group 1 animals received 2×10^8 splenocytes from nontransgenic B10D2 (H-2^d) mice that had been infected intraperitoneally 3 weeks earlier with a recombinant vaccinia virus (HBs-vac) which expresses HBsAg. At the same time, the group 2 animals received the same number of splenocytes from saline-injected, immunologically tolerant transgenic littermates. A third group of transgenic mice (group 3) remained unmanipulated for the entire experiment for baseline comparison (Fig. 1).

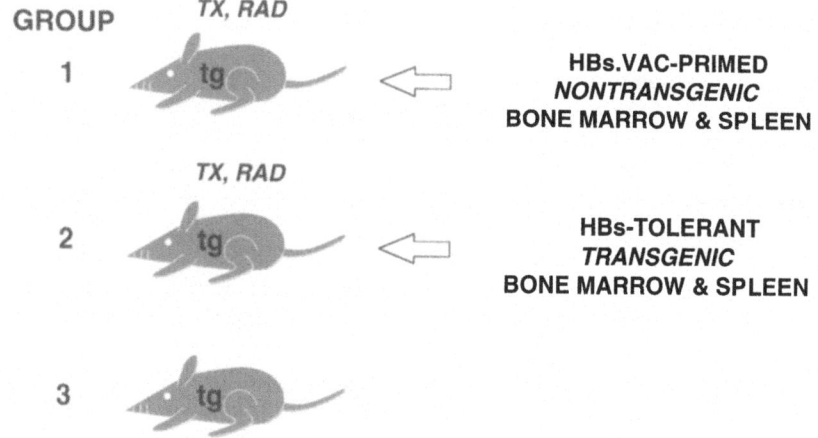

Fig. 1. Experimental design. Group 1 mice were thymectomized, lethally irradiated, and reconstituted with bone marrow and spleen cells from syngeneic nontransgenic donors that had been previously immunized with hepatitis B surface antigen (HBsAg). Similarly treated group 2 transgenic (±g) control animals were reconstituted with bone marrow and spleen cells from immunologically tolerant transgenic donors that had been injected with saline. All results were compared with group 3, unmanipulated, age- and sex-matched transgenic mice

Disease Model

All mice were analyzed for evidence of HBsAg-specific immunity and liver disease before and after BMR and after spleen cell transfer. Serum levels of HBsAg and anti-HBs were assessed by radioimmunoassay (AUSRIA II; Abbott Laboratories, North Chicago, IL, USA) and enzyme-linked immunoassay (AUSAB EIA; Abbott), respectively. Hepatocellular injury was monitored biochemically as serum alanine aminotransferase (sALT) activity. Results were expressed as mean units per liter ± SEM of sALT activity, and differences between experimental and control groups were assessed for statistical significance by Student's t test.

Tumor development was assessed by abdominal palpation and confirmed by autopsy, at which time the number of tumors visible at the surface of each liver was counted and the diameter of each tumor was measured with a millimeter rule. Mice were killed by cervical dislocation. Tissue samples were fixed in 10% zinc-buffered formalin (Anatek, Battle Creek, MI, USA), embedded in paraffin, sectioned (3 μm), and stained with hematoxylin and eosin. The intrahepatic distribution of HBsAg was assessed by the indirect immunoperoxidase method using 3-amino-9-ethyl carbazole (AEC; Shandon-Lipshaw, Pittsburgh, PA, USA) as a coloring substrate. Liver tissue was also snap-frozen in liquid nitrogen and stored at −80°C for molecular analysis; 10 μg of total RNA was extracted from the livers of representative mice and subjected to RNase protection analysis to monitor the expression of CD3γ, CD4, CD8α, F480, and a panel of inflammatory cytokines.

Results and Discussion

The HBsAg-immunized nontransgenic donors displayed high titers of serum anti-HBs antibodies and HBsAg-specific CTLs in their spleens (not shown). In contrast, the transgenic donors were anti-HBs negative and their splenocytes displayed no HBsAg-specific CTL activity, confirming that this lineage is immunologically tolerant to HBsAg. Consistent with the presence of HBsAg-specific B cells and CTLs in the HBsAg primed nontransgenic spleen cell donors, HBsAg disappeared from the serum of the group 1 mice within 3 days after adoptive transfer (Fig. 2A), concomitant with the appearance of anti-HBs (not shown). Furthermore, sALT activity increased in the group 1 recipients within 7 days after adoptive transfer, and fell progressively thereafter (Fig. 2A).

Importantly, sALT activity never returned to baseline in these animals, remaining at least two to three times above normal throughout the experiment and rising further by 17 months after BMR and spleen cell transfer, at which time they were killed. In contrast, serum HBsAg levels did not fall in the group 2 or group 3 mice, in keeping with the absence of anti-HBs in their serum (Fig. 2B). Similarly, sALT activity remained in the normal range in group 2 and group 3 animals throughout most of the study, although it increased late in the experiment in a few of the group 2 radiation controls (Fig. 2B).

The differential sALT activity in the animals was reflected in the gross and histological appearance of their livers at various times following BMR and spleen cell transfer. Group 1 animals analyzed 3 weeks after spleen cell transfer revealed a subacute inflammatory liver disease of variable severity, characterized by a diffuse lymphomononuclear and polymorphonuclear inflammatory infiltrate in the parenchyma

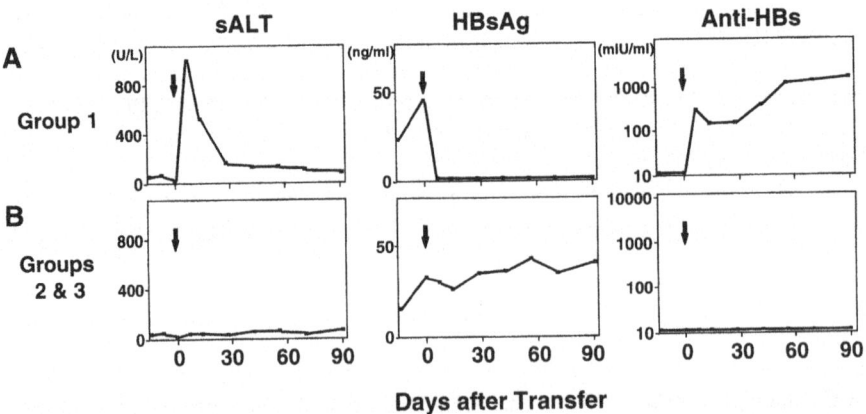

Fig. 2. Kinetics of serum alanine aminotransferase activity (sALT), HBsAg, and anti-HBs in group 1 (A) and groups 2 and 3 (B) hepatitis B virus (HBV) transgenic mice. Adoptive transfer of splenocytes was performed on day 0. Hepatocellular injury was monitored biochemically as sALT. Serum levels of HBsAg and anti-HBs were assessed by radioimmunoassay and enzyme-linked immunoassay, respectively

Table 1. Hepatocarcinogenesis in the context of chronic immune-mediated hepatitis

Group	n	HCC	Age (m)	Final ALT	CTL	HBsAg	αHBs
1	9	8	19	300	+	−	+
2	9	1	20	159	−	+	−
3	23	0	20	80	−	+	−

HCC, hepatocellular carcinoma; ALT, alanine amino transferase; CTL, cytotoxic T lymphocytes; HBsAg, hepatitis B surface antigen; αHBs, anti-HBs

and portal areas, focal hepatocellular necrosis, and dropout and disorganization of the hepatic lobular architecture. Three months later, the mice displayed multiple portal and intralobular inflammatory infiltrates indicative of chronic hepatitis. By 8 months the liver disease in these animals was more severe, displaying hepatocellular apoptosis, dysplasia characterized by abnormal nuclear–cytoplasmic ratios, and pre-neoplastic foci.

Eventually, when the animals were killed 17 months later, they all displayed multiple liver tumors, most of which were greater than 10 mm in diameter (Table 1). The tumors in all but one of these animals displayed histological evidence of well-differentiated, HCC characterized by trabecular cords composed of several layers of well-differentiated neoplastic liver cells covered by a thin endothelial cell lining. The neoplastic hepatocytes contained typical abundant eosinophilic cytoplasm, and the nuclei displayed prominent nucleoli.

In contrast to these findings, the radiation control mice in group 2 displayed much less severe necroinflammatory changes in their livers at the time of autopsy, consistent with their lower sALT levels. Interestingly, two of the nine mice in this group displayed solitary liver tumors up to 10 mm in diameter, one of which was classified histologically as a well-differentiated, trabecular HCC while the other one was a benign adenoma (see Table 1). Not surprisingly, the livers of group 3 mice were grossly and histologically normal, except for the presence of the HBsAg-positive ground glass hepatocytes that are characteristic of this lineage (not shown). The livers of all the animals in this study contained HBsAg+ hepatocytes at the time of autopsy; however, the frequency of such hepatocytes was much lower in group 1 animals (0.01%–15%) than in groups 2 and 3 (30%–60%), reflecting the severity of the underlying inflammatory liver disease (see following).

To characterize the intrahepatic inflammatory response, total hepatic RNA from the nontumorous livers from representative animals from each group was analyzed for the presence of T-cell, macrophage, and cytokine transcripts using an RNase protection assay. The results were compared with total hepatic RNA from a similar group of transgenic mice killed at the peak of a CTL-induced acute hepatitis 3 days after adoptive transfer of 10^7 HBsAg-specific CTL clones. As expected, the highest level of T-cell, macrophage, and cytokine mRNA expression was observed in the acute hepatitis specimens, followed in order by group 1, then group 2, and finally group 3, which represents the baseline for this analysis.

It is noteworthy that interferon-gamma (IFN-γ) mRNA was only detectable in the acutely inflamed livers, suggesting that the quality as well as the magnitude of the inflammatory infiltrate is different in these acute and chronic hepatitis models. These results demonstrate that the inflammatory infiltrate and corresponding cytokine profile (especially tumor necrosis factor-alpha, TNF-α) was more intense in group 1 than group 2 at the time of autopsy. Taken together with the sALT profiles (see Fig. 2), these results indicate that the inflammatory liver disease began much earlier and was more severe in group 1 than in group 2, which probably explains the higher incidence of HCC in the group 1 animals.

To assess the possible contribution of the CTL response to the pathogenesis of the liver disease in all three groups of recipients, spleens harvested at autopsy were examined for HBsAg-specific cytotoxic activity. All the group 1 mice displayed strong HBsAg-specific CTL responses. In contrast, only one of the group 2 control mice showed a weak CTL response, and CTLs were undetectable in all the group 3 controls. The CTL activity of the group 1 spleen cells was inhibited by a monoclonal antibody (Ab) specific for CD8 but not CD4. The antigenic fine specificity of the group 1 CTL response was examined using P815 target cells pulsed with a previously defined L^d-restricted immunodominant CTL epitope located between residues 28 and 39 of hepatitis B virus surface antigen (HBsAg 28–39). CTLs from all but one of the group 1 animals were specific for that epitope, although their cytotoxic activity was generally weaker than that of HBs-vac-primed nontransgenic mouse splenocytes. Splenocytes from the single group 1 mouse whose CTLs did not recognize peptide HBsAg 28–39 responded to a subdominant D^d-restricted CTL epitope located between amino acids 281 and 289 in HBsAg.

These results indicate that CD8-positive HBsAg-specific CTLs were transferred into the group 1 recipients at the beginning of this study and that these CTLs were maintained in these animals for the duration of the experiment. Linear regression analysis was performed to compare the strength of this CTL response with the tumor burden in all three groups of mice. As shown in Fig. 3, we observed a strong relationship between splenic CTL activity and both the number of liver tumors (left) and the size of the largest liver tumor (right) in these animals ($r^2 = 0.62$ and 0.63, respectively). These results were highly statistically significant ($P = 0.0001$ and <0.0001, respectively) using a two-tailed F-test for analysis, suggesting that liver tumor development is directly related to the strength of the CTL response in these animals.

The molecular basis for chronic immune-mediated hepatocellular injury that predisposes to the development of HCC is not yet defined. As we previously observed that the Fas ligand (FasL)/Fas death pathway appears to be operative during a CTL response in acute liver injury [3], monoclonal neutralizing Ab to FasL was introduced into mice treated in the same manner as group 1. The severity of liver disease was diminished to one-third of that at the ALT level 7 days after splenocyte transfer and ALT activity remained low thereafter. Nine months after adoptive transfer, FasL Ab-treated mice demonstrated decreased levels of hepatocyte dysplasia and reduced numbers of preneoplastic foci. Interestingly, the differences in liver histology at the time point between the FasL Ab-untreated group 1 mice and the treated group were reflected by the incidence of the development of liver tumors 15 months after adoptive transfer in each group (100% and 14%, respectively) (Nakamoto et al., in manuscript). Consequently, the results suggest that the FasL-dependent death pathway functions as a procarcinogenic stimulus in the process of hepatocarcinogenesis during chronic hepatitis.

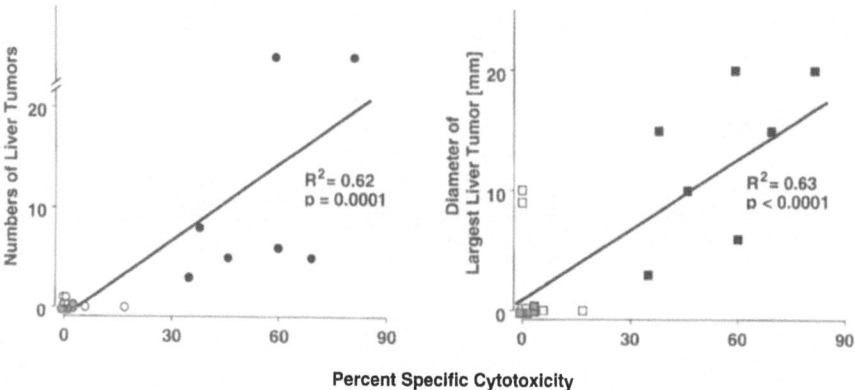

Fig. 3. Correlation between HBsAg-specific cytotoxic activity of splenocytes and number or size of liver tumors in HBV transgenic mice from group 1 (*closed symbols*), group 2 (*open symbols*), and group 3 (*gray symbols*) mice. Tumor development was assessed by abdominal palpation and confirmed by autopsy, at which time the number of tumors visible at the surface of each liver was counted and the diameter of each tumor measured. The HBsAg-specific cytotoxic activity of the splenocytes from each mouse was plotted against the number of visible liver tumors (*left*) and against the diameter of the largest liver tumor (*right*) in the same animal. Linear regression analysis revealed a statistically highly significant relationship between these parameters, with r^2 values of 0.62 and 0.63 reflecting P values of 0.0001 and <0.0001, respectively

The pathogenetic importance of immune-mediated hepatocellular injury in hepatocarcinogenesis during chronic HBV infection is supported by the fact that HCC occurs in the context of necrosis, inflammation, and regeneration (cirrhosis) in other chronic liver diseases including hepatitis C, alcoholism, hemochromatosis, glycogen storage disease, alpha-1-antitrypsin deficiency, and primary biliary cirrhosis. We suggest, therefore, that irrespective of etiology or pathogenesis chronic liver cell injury is a premalignant condition that initiates a cascade of events culminating in the genetic and chromosomal changes necessary to trigger the development of HCC. Chronic necroinflammatory diseases in other organs may also be procarcinogenic in those tissues. For example, several animal models of wounding and inflammation are known to lead to skin cancer. Additionally, in humans continuous oral irritation often leads to oral cancer, inflammatory bowel disease leads to bowel cancer, chronic cystitis leads to bladder cancer, and reflux esophagitis leads to esophageal cancer.

Although the etiopathogenesis of these other diseases is not well defined, the current results demonstrate that chronic, antigen-specific, immune-mediated, FasL-dependent tissue injury is sufficient to initiate and sustain the process of carcinogenesis in the liver. If this is correct, therapeutic interruption of the chronic necroinflammatory disorders should terminate this process and prevent the development of HCC.

References

1. Chisari FV, Klopchin K, Moriyama T, Pasquinelli C, Dunsford HA, Sell S, Pinkert CA, Brinster RL, Palmiter RD (1989) Molecular pathogenesis of hepatocellular carcinoma in hepatitis B virus transgenic mice. Cell 59:1145–1156
2. Nakamoto Y, Guidotti LG, Kuhlen CV, Fowler P, Chisari FV (1998) Immune pathogenesis of hepatocellular carcinoma. J Exp Med 188:341–350
3. Nakamoto Y, Guidotti LG, Pasquetto V, Schreiber RD, Chisari FV (1997) Differential target cell sensitivity to CTL-activated death pathways in hepatitis B virus transgenic mice. J Immunol 158:5692–5697

Induction of Proliferation-Related Signals by Hepatitis C Virus

Naoya Kato, Hideo Yoshida, Motoyuki Otsuka, Yasushi Shiratori, and Masao Omata

Summary. Hepatitis C virus (HCV) causes persistent infection, chronic hepatitis, cirrhosis, and hepatocellular carcinoma. To explore the influence of HCV infection on hepatocytes, the effects of HCV proteins on intracellular signal transduction pathways were investigated. The effects of seven HCV proteins (core, nonstructural [NS]2, NS3, NS4A, NS4B, NS5A, and NS5B) on cyclic AMP response element (CRE)-, serum response element (SRE)-, nuclear factor (NF)-κB-, activator protein (AP)-1-, serum response factor (SRF)-, and p53-associated pathways were investigated by use of a reporter assay. The activation of signals by HCV proteins was examined using a reporter plasmid with an interleukin (IL)-8 or $p21^{wafl}$ promoter. The possible mechanisms by which HCV proteins activate these pathways were investigated. Among the seven HCV proteins investigated, core protein had the strongest influence on intracellular signaling, especially SRE-, AP-1-, NF-κB-, and p53-associated pathways. Core protein activated IL-8 promoter through NF-κB and AP-1 and activated p21 promoter having a p53-binding site. Core protein activated the NF-κB pathway mainly through IKKβ and tumor necrosis factor receptor-associated factor 2/6 and augmented p53 function by increasing both p53–DNA binding affinity and transcriptional ability itself. Direct interaction between core protein and the C-terminus of p53 was detected. In addition, the interaction between core protein and human TBP-associated factor $_{II}28$, a component of the transcriptional factor complex, was also demonstrated. Core protein may directly promote cell proliferation and induce an inflammatory reaction by activating SRE-, AP-1-, and NF-κB-associated pathways. On the other hand, core protein could enhance p53 function. These opposing functions may result in exquisitely balancing the proliferation of hepatocytes infected with HCV.

Key words. AP-1, HCV core, IL-8, Intracellular signal transduction, MAPK cascade, NF-κB, $p21^{wafl}$, p53, SRE, Transcriptional factor

Department of Gastroenterology, Graduate School of Medicine, University of Tokyo, 7-3-1 Hongo, Bunkyo-ku, Tokyo 113-8655, Japan

Introduction

Hepatitis C virus (HCV), a positive-stranded RNA virus with a genome size of approximately 10 kb [1], is a major causative agent of chronic hepatitis, cirrhosis, and hepatocellular carcinoma (HCC) [2]. HCV infection is widespread throughout the world. The World Health Organization reported that more than 170 million people were infected with HCV and were at risk for developing liver cirrhosis and/or HCC [3].

The HCV genome contains a large open reading frame encoding a polyprotein precursor of about 3000 amino acids (aa) and an untranslated region at the 5'- and 3'-ends of the genome. This polyprotein is processed into at least ten proteins: four structural proteins (core, E1, E2, and p7) and six nonstructural (NS) proteins (NS2, NS3, NS4A, NS4B, NS5A, and NS5B) [4,5]. To elucidate the effect of HCV infection on hepatocytes, we analyzed the induction of intracellular signal transduction pathways using seven HCV proteins (core, NS2, NS3, NS4A, NS4B, NS5A, and NS5B). Because an efficient cell culture system permissive of HCV infection and replication has not been established, we adopted a transient transfection system with HCV protein expression vectors and six reporter vectors containing the luciferase gene driven by the following well-defined inducible cis-enhancer elements: the cyclic AMP response element (CRE), the serum response element (SRE) [6], and the binding sites for nuclear factor κB (NF-κB) [7], activator protein 1 (AP-1) [8], serum response factor (SRF) [6], and p53 [9]. In addition to these reporter plasmids with synthetic promoters, the activation of intracellular signals by HCV proteins was examined with a reporter plasmid having a natural interleukin (IL)-8 or p21^{waf1} promoter upstream from a firefly luciferase gene. Moreover, the possible mechanisms of activating these pathways by HCV proteins were investigated.

Materials and Methods

Cell Lines

Human cervical carcinoma cells (HeLa), human hepatoblastoma cells (HepG2), human osteosarcoma cells (SAOS-2), and monkey kidney cells (COS-7) were obtained from the Riken cell bank (Tsukuba, Japan).

Mammalian Expression Plasmids for HCV Proteins

The mammalian expression plasmids pCXN2-core, NS2, NS3, NS4A, NS4B, NS5A, and NS5B containing the respective HCV genomic regions, driven by a β-actin-based CAG promoter [10], were constructed as described previously [11]. pCXN2-core173 and pCXN2-core151, which express C-terminal 18- and 40-aa truncated core proteins, respectively, were also prepared as described previously [12]. pCXN2 (kindly provided by J. Miyazaki, Osaka University, Osaka, Japan) was used as a control.

Intracellular Signal Transduction Pathway Reporter Plasmids

A series of vectors containing the *Photinus pyralis* (firefly) luciferase reporter gene driven by an inducible *cis*-enhancer element were utilized as reporter plasmids. Expression of the firefly luciferase gene was controlled by a synthetic promoter containing CRE (pCRE-Luc) (Stratagene, La Jolla, CA, USA), SRE (pSRE-Luc) (Stratagene), the binding site for NF-κB (pNF-κB-Luc) (Stratagene), the binding site for AP-1 (pAP1-Luc) (Stratagene), the binding site for SRF (pSRF-Luc) (Stratagene), and the binding site for p53 (pG13-luc) [13] (kindly provided by B. Vogelstein, Johns Hopkins University, Baltimore, MD, USA).

In addition to these reporter plasmids with synthetic promoters, a luciferase reporter plasmid having a 5'-flanking region of the human IL-8 gene spanning from −133 bp to +44 bp (pIL-8 luc) (kindly provided by K. Matsushima, University of Tokyo, Tokyo, Japan) was utilized [14]. The IL-8 promoter region contains the following three *cis* elements important for induction of IL-8 gene expression: AP-1, NF-IL-6, and NF-κB binding sites. An additional three plasmids having mutations in the IL-8 AP-1, NF-IL-6, and NF-κB binding sites, respectively, were also utilized [15]. Moreover, WWP-luc (kindly provided by B. Vogelstein), a reporter plasmid having a 2.4-kb genomic fragment of the human p21 promoter region, which contains the p53-binding site, was also utilized [16]. Transfection efficiency was monitored through the cotransfection of pRL-TK (Promega, Madison, WI, USA), a control plasmid expressing *Renilla reniformis* (sea pansy) luciferase driven by the herpes simplex virus thymidine kinase.

pCXN2-p53, a human wild-type p53 expression vector, and pCXN2-Gal4BD-p53, a mammalian expression vector for Gal-4 DNA-binding domain fused with p53 protein, were constructed as described previously [17]. For immunoprecipitation of core and TFIID, human TATA binding protein (hTBP) and five human TBP-associated factor (hTAFs) expression plasmids were also constructed as described previously [17]: N-terminal HA-tagged hTBP expression plasmid, pCXN2-HA-hTBP; a hTAF$_{II}$32/31 expression plasmid, pCXN2-hTAF$_{II}$32/31; and N-terminal HA-tagged hTAF$_{II}$70, hTAF$_{II}$28, hTAF$_{II}$20, and hTAF$_{II}$18 expression plasmids, pCXN2-hTAF$_{II}$70, pCXN2-hTAF$_{II}$28, pCXN2-hTAF$_{II}$20, and pCXN2-hTAF$_{II}$18, respectively.

To elucidate the mechanism by which the HCV core protein affects the NF-κB pathway, the expression vectors for catalytically inactive IKKα [pIKKα$_{(K44A)}$] [18] and IKKβ [pIKKβ$_{(K44A)}$] [19] (kindly provided by D.V. Goeddel, Tularik, South San Francisco, CA, USA) were utilized. In addition, the expression vectors for the dominant negative form of tumor necrosis factor (TNF) receptor-associated factor 2 (TRAF-2) [pTRAF-2 (87–501)] [20] and TRAF-6 [pTRAF-6 (289–522)] [21] (kindly provided by D.V. Goeddel) were utilized.

Construction of HeLa Cells Induced to Express Core Protein

HeLa cells induced to express HCV core protein, termed HeTOC, were generated with use of a tetracycline-regulated gene expression system (Tet-off gene expression system; Clontech, Palo Alto, CA, USA) as described previously [12].

Transfection

The transfection of plasmids was performed using Effectene (Qiagen, Hilden, Germany) or FuGene6 (Roche Molecular Biochemicals, Mannheim, Germany) transfection reagent. To assess whether HCV proteins activated intracellular signal transduction pathways, the transfection complexes containing a firefly luciferase reporter plasmid, pRL-TK, and pCXN2/pCXN2-HCV were added to HeLa or HepG2 cells as recommended by the manufacturer's instructions.

Luciferase Assay

Cells were harvested 36–48h after transfection, and luciferase assays were carried out with the dual luciferase assay (PicaGene dual seapansy system; Toyo Ink, Tokyo, Japan). Firefly luciferase and seapansy luciferase activities were measured as relative light units with a luminometer (Lumat LB9507; EG&G Berthold, Bad Wildbad, Germany). Firefly luciferase activity was then normalized for transfection efficiency based on seapansy luciferase activity.

Indirect Immunofluorescence Staining of Core Proteins

COS-7 cells were transfected with pCXN2-core, pCXN2-core173, or pCXN2-core151. At 48h after transfection, COS-7 cells were fixed with 3.7% formaldehyde, blocked with phosphate-buffered saline containing 2% normal rabbit serum, and then incubated with mouse anticore antigen IgG fraction (anti-core Ab) (1:500) (Austral Biologicals, San Ramon, CA, USA). Cells were then incubated with fluorescein isothiocyanate-conjugated rabbit antimouse IgG Ab (1:40) (Dako, Carpinteria, CA, USA), and observed by microscope using an epifluorescent attachment (AX80; Olympus, Tokyo, Japan).

Electrophoretic Mobility Shift Assay

To examine NF-κB binding to the κB site, an electrophoretic mobility shift assay (EMSA) was performed using a gel shift assay system (Promega) according to the manufacturer's instructions. HeTOC cells were cultured in medium with or without 1mg/ml doxycycline. After 48h, the cells were harvested and their nuclear extracts were prepared according to mininuclear extraction methods [22]. A synthetic double-stranded oligonucleotide having a κB site (5'-AGTTGAGGGGACTTTC-CCAGGC-3') was end-labeled with [^{32}P]ATP and incubated with nuclear extracts. For competition, an unlabeled competitor oligonucleotide or an unlabeled noncompetitor oligonucleotide (5'-GATCGAACTGACCGCCCGCGGCCCGT-3') was added to the reaction mixture in 100-fold excess over the labeled probe to examine the binding specificity.

Similarly, EMSA was performed to examine the p53 binding to its responsive element using a p53 NuShift kit (Geneka, Montréal, Canada) according to the manufacturer's instructions. SAOS-2 cells were transfected with pCXN2-p53 and pCXN2/pCXN2-core, and nuclear extracts were prepared as already described. To visualize p53–DNA complexes, p53-activating monoclonal Ab Pab421 (Calbiochem,

La Jolla, CA, USA) was added to all mixtures. The specificity of p53 in this assay was tested by adding 100-fold excess of either wild-type or mutant competitor (5′-GGATCGCCCCGGGCATGTC-3′), as well as the supershift caused by addition of the anti-p53 Ab (pAb1801; Calbiochem).

Immunoblotting

To determine the expression levels of p53 and p21, immunobotting was performed using rabbit anti-p53 and goat anti-p21 IgG polyclonal Abs for the first Ab (Santa Cruz Biotechnology, Santa Cruz, CA, USA). For the second antibody, antirabbit Ab (Amersham Pharmacia Biotech, Piscataway, NJ, USA) and antigoat Ab (Santa Cruz Biotechnology) conjugated with horseradish peroxidase were used. The bound antigen was detected by ECL-plus Western blotting detection system (Amersham Pharmacia Biotech).

Glutathione S-Transferase Binding Assay

pGEX4T1-p53, p53N, and p53C, which express 1–393, 1–71, or 72–393 aa of p53, respectively, fused to glutathione S-transferase (GST) in *Escherichia coli* were constructed as previously described [17]. These constructs were expressed in *E. coli* and purified with glutathione-sepharose 4B beads as specified by the manufacturer (Amersham Pharmacia Biotech). pCXN2-core was transfected into COS-7 cells, and the cell lysates were prepared in immunoprecipitation (IP) buffer containing 50 mM Tris-HCl (pH 7.5), 150 mM NaCl, 0.1% NP-40, 1 mM ethylenediaminetetraacetic acid (EDTA), 0.25% gelatin, 0.02% sodium azide, 100 μg/ml phenylmethylsulfonyl fluoride, and 1 μg/ml aprotinin. Each fusion protein was bound to glutathione sepharose and incubated with cell lysates, and the bound proteins were separated by sodium dodecyl sulfate-polyacrylamide gel electrophoresis (SDS-PAGE) and immunoblotted by anti-core Ab.

Immunoprecipitation

COS-7 cells were transfected with pCXN2-core together with hTAF$_{II}$ vectors. Cell extracts were prepared with IP buffer and precipitated by incubation with anti-HA Ab or anti-core Ab followed by addition of protein A sepharose (Amersham Pharmacia Biotech). The precipitated proteins were revealed by immunoblotting using anti-core Ab or anti-hTAF$_{II}$31/32 Ab (Santa Cruz Biotechnology).

Results

Activation of Signal Transduction Pathways by HCV Proteins

HCV protein activation three times greater than that of the control was defined as significant. Core protein significantly activated the NF-κB-associated signal in HeLa cells at a value 6.0 ± 3.4 (mean ± SD) times higher than the control (Fig. 1). Core protein also activated SRE-associated (4.5 ± 3.1 times) and AP-1-associated (3.3 ± 1.4 times)

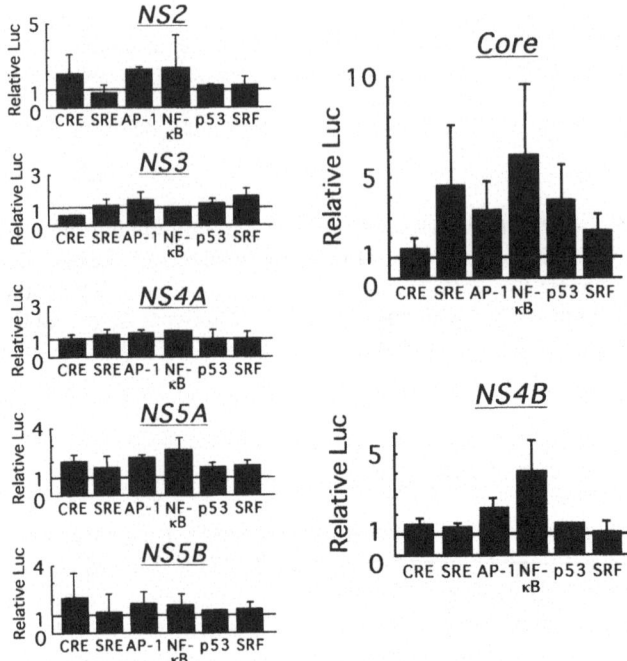

Fig. 1. Luciferase assays showing activation of each *cis*-enhancer element by hepatitis C virus (HCV) proteins in HeLa cells (cyclic AMP response element [CRE], serum response element [SRE], activator protein [AP]-1, nuclear factor κB [NF-κB], and serum response factor [SRF]) or SAOS-2 cells (p53). Luciferase activity was normalized by taking the activity of pCXN2-transfected cell lysate as 1 (relative luciferase activity). Results are expressed as the mean (*bar*) ± SD (*line*) of at least three experiments. *Relative Luc*, relative luciferase activity

signals (Fig. 1). Core protein activated the NF-κB- and SRE-associated signals but not the AP-1-associated signal in HepG2 cells (data not shown). Moreover, core protein activated the p53-responsive element in SAOS-2 in the presence of exogenously expressed p53 (Fig. 1). Besides core protein, NS4B protein activated the NF-κB-associated signal at a value 4.0 ± 1.5 times higher than the control (Fig. 1).

Activation of the IL-8 Promoter Through NF-κB and AP-1 by Core Protein

As core protein significantly activated NF-κB- and AP-1-associated signals, we examined the effect of core protein on IL-8 promoter. Core protein enhanced luciferase activity (7.0 ± 4.3 times higher than the control) in HeLa cells transfected with pIL-8 luc (Fig. 2). Because this promoter region contains three important *cis* elements for IL-8 gene transcription, the AP-1, NF-IL-6, and NF-κB binding sites, the relative contribution of each element to core-induced IL-8 gene transcription was examined. Mutation of the NF-IL-6 binding site reduced core-induced enhancement of luciferase activity by only 21%, whereas mutation of the AP-1 binding site decreased this core-induced enhancement by more than 70% (Fig. 2). Moreover, mutation of the NF-κB

Fig. 2. Activation of IL-8 gene transcription by core protein through activation of NF-κB and AP-1 in HeLa cells. Luciferase activities were measured and expressed as described in the legend for Fig. 1. *mut*, mutant; *luc*, luciferase; *Relative luc*, relative luciferase activity

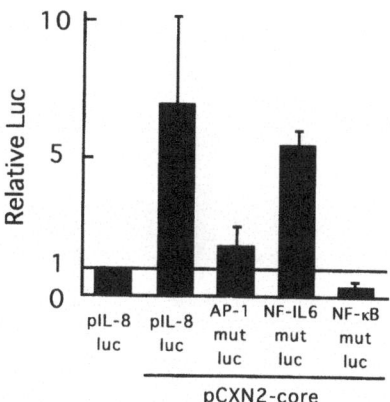

binding site completely abrogated the enhancement induced by core protein (Fig. 2). These results suggest that the NF-κB binding site and, to a lesser degree, the AP-1 binding site are involved in core-induced IL-8 gene transcription.

Enhancement of p21 Promoter Activity and Induction of p21 Protein by Core Protein

Because core protein activated p53-responsive element, WWP-luc containing the promoter of p21, a well-known p53 target gene, was cotransfected with a fixed level of p53 expression plasmid and each of seven HCV protein expression plasmids into p53-null SAOS-2 cells. Thirty-six hours later, the cells were assayed for luciferase activity. The relative firefly luciferase activity of WWP-luc and pCXN2-core transfected cell lysate was 4.3 ± 0.1 (mean $\pm$ SD) times higher than the control (Fig. 3A). The other HCV proteins, however, did not influence p21 promoter activity.

Because p21 promoter activity was enhanced by core protein in the luciferase assay system, cellular p21 protein expression was examined. Endogenous levels of p21 protein were determined by Western blot analysis using the SAOS-2 cell lysates. The levels of endogenous p21 expression increased in relation to the amount of core protein (Fig. 3B).

Core Protein Enhances NF-κB–DNA Binding Activity

To examine whether core protein enhances NF-κB–DNA binding, EMSA was performed using the nuclear extracts of HeTOC cells, which were induced to express core protein. NF-κB–DNA binding activity was enhanced in HeTOC cells expressing core protein (about 2.9 times) as compared with that in HeTOC cells without expressing core protein under the existence of doxycycline. This NF-κB–DNA binding activity was ablated by an excess of unlabeled competitor but not by an excess of unlabeled noncompetitor, showing its specificity (Fig. 4). Addition of an Ab directed against p65 or p50 (Santa Cruz Biotechnology) generated a supershifted band, suggesting that this NF-κB–DNA complex contained p65 and p50 (data not shown).

A

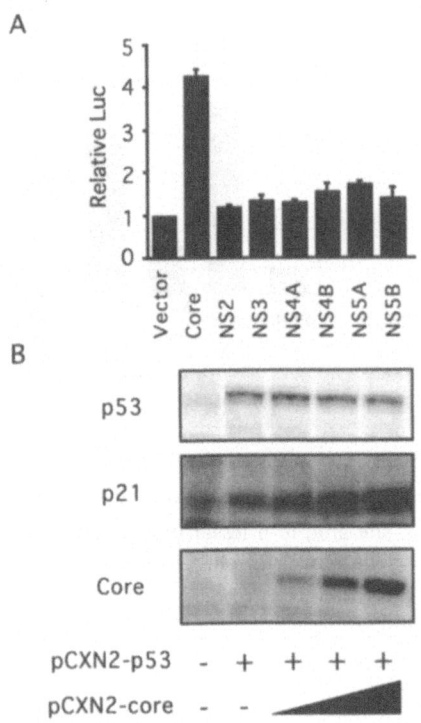

B

p53

p21

Core

pCXN2-p53   - + + + +

pCXN2-core   - - ◢ ◢ ◢

Fig. 3. A The effect of HCV proteins on p21 promoter activity. SAOS-2 cells were cotransfected with WWP-luc, pRL-TK, and various expression plasmids of HCV proteins with pCXN2-p53. Luciferase activities (*Relative luc*) were measured and expressed as described in the legend for Fig. 1. **B** Increased induction of endogenous p21 protein by core protein. SAOS-2 cells were transfected as already described. After 36 h luciferase activity was measured and the rest of the cell lysate was normalized for protein concentration and used for immunoblotting of p53, p21, and core protein with each antibody, respectively. Representative result of immunoblotting is indicated

HeLa-core

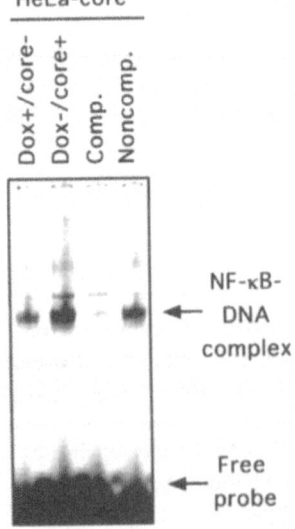

NF-κB-
DNA
complex

Free
probe

Fig. 4. Enhancement of NF-κB DNA-binding activity by core protein. HeTOC cells were cultured in medium with or without 1 mg/ml doxycycline. After 48 h, nuclear extracts were prepared and assayed for the NF-κB–DNA-binding activity by electrophoretic mobility shift assay (EMSA). The following were added to the reaction: *lane Dox+/core–*, nuclear extracts from HeTOC cells with doxycycline (no expression of core protein); *lane Dox–/core+*, nuclear extracts from HeTOC cells without doxycycline (expressing core protein). *Lane Comp.* and *lane Noncomp.* contained the same nuclear extracts as lane Dox–/core+, except either excess unlabeled competitor oligonucleotide probe or excess unlabeled noncompetitor oligonucleotide probe was added for competition, respectively

Core Protein Activates NF-κB Pathway Through IKKβ

To examine whether activation of the NF-κB pathway by core protein is transduced through IKKα/IKKβ, HeLa cells were cotransfected with pCXN2/pCXN2-core, pNF-κB-Luc, and pIKKα$_{(K44A)}$/pIKKβ$_{(K44A)}$. Expression of IKKβ$_{(K44A)}$ significantly reduced core-induced NF-κB activation to about one-tenth, whereas expression of IKKα$_{(K44A)}$ reduced the activation to about two-fifths (Fig. 5A). To confirm the participation of IKKβ in activation of the NF-κB pathway by core protein, 5 mM acetyl salicylic acid, an IKKβ-specific inhibitor [23], was added to the HeLa cells transfected with pCXN2/pCXN2-core and pNF-κB-Luc. Activation of the pathway by core protein was significantly inhibited by acetylsalicylic acid but not by indomethacin, a cyclooxygenase inhibitor (Fig. 5B). These results suggest that core protein activates the NF-κB pathway through IKK, especially IKKβ.

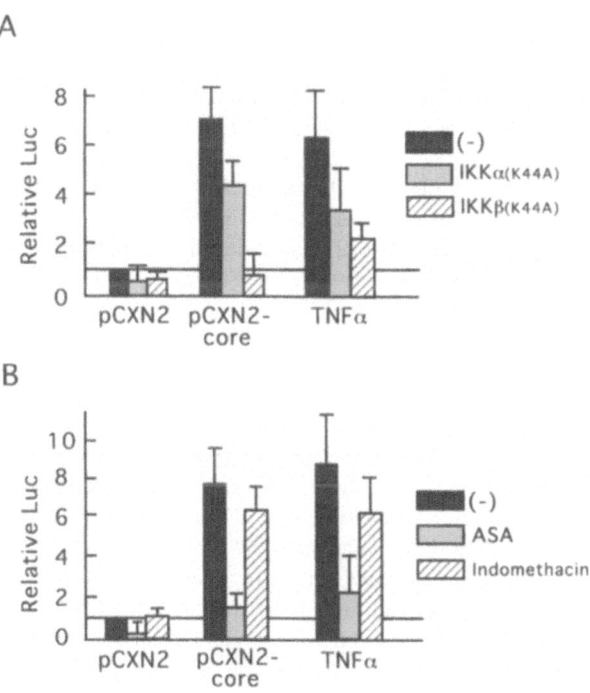

Fig. 5. A Catalytically inactive IKK reduced NF-κB activation by core protein. HeLa Cells were transfected with pCXN2/pCXN2-core, pIKKα$_{(K44A)}$/pIKKβ$_{(K44A)}$, and pNF-κB-Luc. Luciferase activities (*Relative luc*) were measured and expressed as described in the legend for Fig. 1. Tumor necrosis factor (TNF)-α (20 ng/ml) was added 6 h before harvest to function as an inducer of the pathway (positive control). **B** Acetylsalicylic acid (*ASA*) reduced NF-κB activation by core protein. ASA, an IKKβ-specific inhibitor, was added to the medium of HeLa cells transfected with pCXN2/pCXN2-core and pNF-κB-Luc at a concentration of 5 mM. Luciferase activities were measured and expressed as described in the legend for Fig. 1

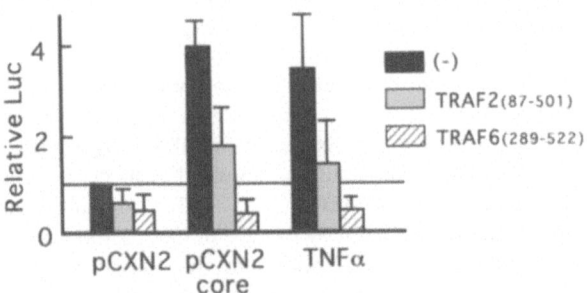

Fig. 6. Dominant negative tumor necrosis factor receptor-associated factor (TRAF)2/6 reduced NF-κB activation by core protein. HeLa cells were transfected with pCXN2/pCXN2-core, pTRAF2(87–501)/pTRAF6(289–522), and pNF-κB-Luc. Luciferase activities were measured and expressed as described in the legend for Fig. 1. TNF-α (20 ng/ml) was added 6 h before harvest to function as an inducer of the pathway (positive control). *Relative Luc*, relative luciferase activity

Core Protein Activates NF-κB Pathway Through TRAF2/6

To examine whether activation of the NF-κB pathway by core protein was transduced through TRAF2/6, HeLa cells were cotransfected with pCXN2/pCXN2-core, pNF-κB-Luc, and pTRAF2(87–501)/pTRAF6(289–522). Expression of the dominant negative form of TRAF-2 (aa 87–501) significantly reduced core-induced NF-κB activation to about two-fifths in HeLa cells (Fig. 6). Expression of the dominant negative form of TRAF-6 (aa 289–522) also reduced core-induced NF-κB activation (Fig. 6). These results suggest that core protein activates the NF-κB pathway through TRAF2/6.

Mapping the Region of Core Protein Responsible for Activation of the NF-κB Pathway

To determine the region of core protein responsible for activation of the NF-κB pathway, HeLa cells were transfected with full-length or each deletion mutant form of core-expressing plasmid in combination with pNF-κB-Luc (Fig. 7). None of the deletion mutant forms of core protein activated the NF-κB pathway, whereas full-length core protein did activate the pathway, suggesting that C-terminus (aa 174–191) of the core protein may play an important role in activation of the NF-κB pathway.

Then, subcellular localization of full-length and deletion mutant forms of core protein was examined by indirect immunofluorescence assay. Full-length core protein (aa 1–191) was located diffusely in the cytoplasm, in contrast to the perinuclear localization of the C-terminal 18-aa-deleted core protein (aa 1–173) and the nuclear localization of the C-terminal 40-aa-deleted core protein (aa 1–151) (Fig. 7).

Core Protein Enhances p53–DNA Binding Activity

To elucidate how core protein enhances the transcription of p53-responsive elements, modulation of p53–DNA binding activity by core protein was investigated. EMSA was

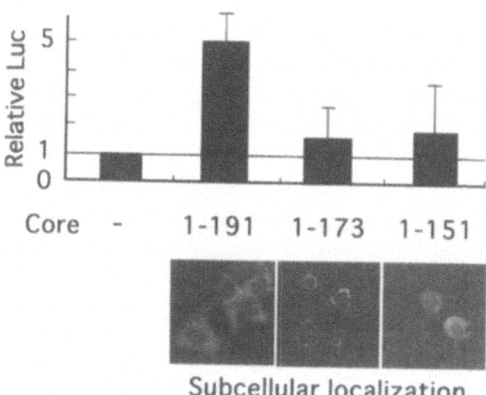

Subcellular localization

Fig. 7. Mapping the region of core protein responsible for activation of the NF-κB pathway and subcellular localization of full-length and deletion mutant forms of core protein. HeLa cells were transfected with pCXN2/pCXN2-core/pCXN2-core173/pCXN2-core151 and reporter plasmids. Luciferase activities were measured and expressed as described in the legend for Fig. 1. Subcellular localization of full-length and deletion mutant forms of the core protein was also analyzed by indirect immunofluorescence assay. COS-7 cells were transfected with three types of core protein expression plasmid (pCXN2-core, pCXN2-core173, and pCXN2-core151). After 48 h, cells were fixed and incubated with anticore Ab and then stained with fluorescein isothiocyanateconjugated rabbit antimouse IgG Ab. *Relative Luc*, relative luciferase activity

performed using the nuclear extracts of SAOS-2 cells transfected with pCXN2-p53 and pCXN2/pCXN2-core. P53–DNA binding activity was enhanced in core protein-expressing cells compared with that in empty vector transfected cells (Fig. 8).

Core Protein Enhances p53 Transcriptional Ability

To elucidate whether core protein modulates p53 transcriptional ability itself irrespective of its DNA-binding activity, pCXN2-Gal4BD-p53, a Gal4DNA-binding domain–p53 hybrid protein expression vector, and pFR-luc, Gal4DNA-binding sequences linked with the luciferase gene, were used. Because DNA binding of Gal4BD–p53 was mediated by the Gal4 sequence in this system, luciferase activity reflected the transcriptional ability of p53 itself. Various amounts of pCXN2-core with fixed amounts of pCXN2-Gal4BD-p53 and pFR-luc were cotransfected into SAOS-2 cells. As shown in Fig. 9, core protein enhanced p53 transcriptional ability. These results showed that the enhancement of p53 function by core protein might depend not only on augmentation of the p53–DNA binding affinity but also on enhancement of the transcriptional ability of p53 itself.

Core Protein Binds p53 in vitro

To clarify the molecular mechanism underlying the enhancement of p53–DNA binding affinity and transcriptional ability of p53 by core protein, direct interaction

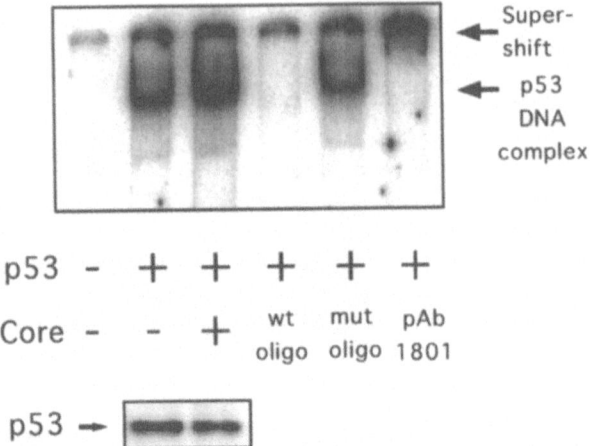

Fig. 8. Core protein enhances p53–DNA binding affinity. SAOS-2 cells were transfected with pCXN2-p53 and pCXN2/pCXN2-core. After 36 h, nuclear extracts were prepared and assayed for p53–DNA binding affinity by EMSA. The following were added to the reaction: *lane 1*, nuclear extract from cells without transfection as negative control; *lane 2*, nuclear extract from cells transfected with pCXN2-p53 and pCXN2; *lane 3*, nuclear extract from cells transfected with pCXN2-p53 and pCXN2-core; *lanes 4* and *5*, the same nuclear extracts as lane 2, except that either excess unlabeled mutant oligonucleotide (*mut oligo*) probe or excess p53 consensus binding oligonucleotide probe was added for competition, respectively; *lane 6*, the same nuclear extracts as lane 2 except that anti-p53 N-terminal Ab (*pAb1801*) was added for supershift analysis. All reactions contained Pab421. Nuclear extracts from the cells were immunoblotted with anti-p53 Ab

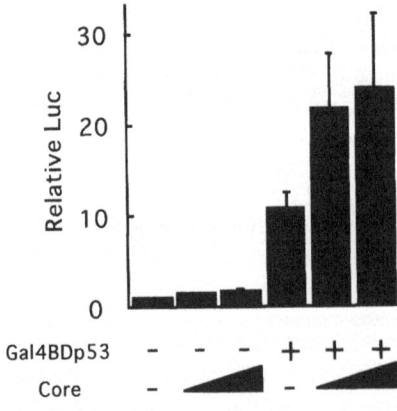

Fig. 9. Core protein enhances the transcriptional ability of p53 itself. SAOS-2 cells were transfected with pFR-luc, pRL-TK, pCXN2-Gal4DBD-p53, and increasing amounts of pCXN2-core. Luciferase activities (*Relative Luc*) are expressed as described in the legend of Fig. 1

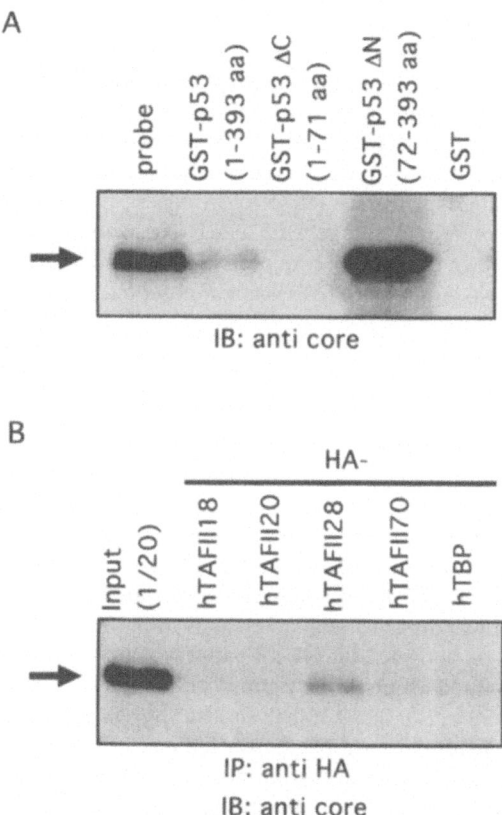

Fig. 10. A Interaction of core protein with p53 in the glutathione S-transferase (GST) fusion protein assay. COS-7 cells were transfected with pCXN2-core; 36 h after transfection, the lysates were prepared by immunoprecipitation buffer. The lysates were mixed with *GST, GST-p53, GST-p53N*, or *GST-p53C* fusion proteins bound to glutathione sepharose 4B beads, respectively, as indicated. Bound proteins were then analyzed by SDS-PAGE followed by immunoblotting with anti-core Ab. Core protein is shown by an *arrow*. **B** Analysis of the interaction of core protein with TFIID. COS-7 cells were transfected with pCXN2-core together with an HA-tagged vector expressing TFIIDs as indicated *above* the lanes. Cell extracts were precipitated with anti-HA Ab and the precipitated proteins were revealed by immunoblotting using anti-core Ab. The coprecipitated core protein is shown by an *arrow*. For input, 1/20 of the cell extracts was used

between core protein and p53 was examined by GST binding assay. As shown in Fig. 10A, core protein bound to GST–p53 but not to GST.

To determine which region of p53 was required for interaction with core protein, N-terminal deletion (p53C; aa 72–393) and C-terminal truncation (p53N; aa 1–71) mutants of p53 were fused to GST protein and used for the GST-binding assay. As shown in Fig. 10A, core protein bound to GST–p53C but not to GST–p53N, suggesting that aa 72–393 of p53 was important for core protein binding.

Interaction Between Core Protein and hTAF$_{II}$28

Although core protein itself had no transcriptional ability (data not shown), it was possible that core protein might act as a transcriptional coactivator. Prior studies have established that p53 interacts with hTAF$_{II}$32/31, hTAF$_{II}$70, and TBP in a transcriptional initiation complex. Thus, to investigate the possible contribution of TFIID composed of TBP and hTAFs to the enhancement of p53 transcriptional ability by core protein, the ability of core protein to interact with TFIID was examined by coimmunoprecipitation assay. As shown in Fig. 10B, core protein interacted with hTAF$_{II}$28. Interaction between core protein and other hTAFs was not observed.

Discussion

The influence of HCV proteins on cellular signal transduction pathways has yet to be fully understood. In this study, we determined the effect of almost all HCV proteins (except envelope proteins E1, E2, and p7) on six major signal transduction pathways in an attempt to clarify the influence of HCV infection on hepatocytes. Among seven HCV proteins, core protein had the strongest influence on intracellular signaling and activated SRE-, NF-κB-, AP-1-, and p53-associated signals.

One of the major findings is the marked activation of the NF-κB pathway by core protein. Core protein activated the NF-κB pathway mainly through IKKβ. Recently, it was shown that only IKKβ phosphorylation contributes to IKK activation by proinflammatory cytokines [24]. Genetic analysis also showed that disruption of the IKKβ locus results in a major defect in IKK activation and IκB degradation in response to proinflammatory stimuli, whereas disruption of the IKKα locus has no effect on either event [25,26]. Furthermore, the dominant negative forms of TRAF2/6 significantly blocked activation of the NF-κB pathway by core protein. TRAF-2 is thought to be a common mediator of TNF receptor (TNFR) and CD 40 signaling [20], whereas TRAF-6 is thought to be a signal transducer for IL-1 [21]. Because CD 40 is not expressed in HeLa and HepG2 cells (data not shown), TNFR or IL-1 signaling is the temporary candidate for the target of core protein. Although TRAF-6 is thought be involved in IL-1 signaling, previous and present studies have shown that the dominant negative form of TRAF-6 suppresses the activation of TNF-α activation of NF-κB [27]. These data may imply that core protein mimics the proinflammatory cytokine activation of the NF-κB pathway, especially TNFR signaling through TRAF2/6. Deletion analysis of the core protein showed cytoplasmic localization was important for NF-κB activation, supporting our finding that core protein activates NF-κB in the cytoplasm through TRAF2/6.

The NF-κB pathway plays an important role in cellular response to proinflammatory cytokines such as TNF-α and IL-1 and induces an inflammatory response by the upregulation of many cytokines, including IL-1, -2, -6, -8, -12, and TNF-α [7]. In this study, core protein was shown to mimic proinflammatory cytokine activation of the NF-κB pathway, especially TNFR signaling. Therefore, it is quite conceivable that core protein could induce an inflammatory response and cause hepatitis. Core protein actually activated a natural IL-8 promoter, mainly through NF-κB. IL-8 is a member of the superfamily of C-X-C chemokines and is a major neutrophil recruiting agent.

In fact, it was reported that the level of IL-8 expression in the liver is strongly associated with the severity of inflammation in patients with chronic hepatitis C [28]. Moreover, serum or intrahepatic expression of IL-1β, -2, -6, and -8 and TNF-α are elevated from two to ten times higher in patients with active chronic hepatitis C than those of a control group and are reduced after eradication of the virus by interferon treatment [28–30]. Although the host immune response caused by cytotoxic T lymphocytes is believed to play a pivotal role in the pathogenesis of C-viral hepatitis [31], our findings suggest that core protein directly induces hepatitis through inflammatory cytokine production. In addition, HCV was reported to infect and replicate not only in the liver but also in the peripheral blood mononuclear cells [32]. Thus, core protein may modulate the immune reaction induced by these HCV-infected mononuclear cells.

The NF-κB-associated signal is recognized as a survival signal. NF-κB regulates a large number of genes involved in cell activation and growth control and is thus thought to be responsible for cell growth and cell transformation [7]. Moreover, NF-κB has an antiapoptotic function against TNF-α-induced cell death [33–35]. Core protein was reported to have an antiapoptotic function against TNF-α-induced cell death [36,37]. In addition to the NF-κB pathway, a significant activation of AP-1- and SRE-associated signals was observed by core protein in this study. One of the cis-enhancer elements, AP-1, is a heterodimeric complex containing products of the jun and fos oncogene families, and mediates signals from growth factors, inflammatory peptides, oncogenes, and tumor promoters, resulting in cell proliferation [8]. Another cis-enhancer element, SRE, is present in the upstream sequence of a number of immediate early genes such as c-fos, and binds the complex of p67^{SRF} and an Ets family protein such as Elk-1 to activate transcription [6]. Phosphorylation of Elk-1 by activated mitogen-activated protein kinase (MAPK) is essential for transcriptional activity of Elk-1 [6]. This MAPK cascade is considered to be involved in the regulation of cell proliferation [38]. Thus, the activation of three cascades (NF-κB-, AP-1-, and SRE-associated signals) by core protein may enhance cell proliferation. In fact, recent studies have revealed that core protein transformed primary mouse and rat fibroblasts in cooperation with H-ras [39] and induced HCC in transgenic mice [40]. Core protein may contribute to cell proliferation and production of inflammatory cytokines through three major pathways (NF-κB, AP-1, and SRE): the former is related to hepatic neoplasia and the latter to inflammation of the liver.

In this study, we also showed that core protein augmented the promoter activity and the expression of p21. In the absence of exogenous p53 in p53-null SAOS-2 cells, however, enhancement of p21 promoter activity by the core protein was not observed, suggesting p53-dependent enhancement of p21 promoter activity by core protein. Several viral proteins have been reported to enhance p53 function [41–44]. These proteins are related to viral replication and act as a promiscuous trans-activator of a variety of viral and cellular genes. Therefore, the mechanism of this enhancement is explained partly by the increase in p53 protein level during the transcription process. In this study, however, we showed that core protein enhanced p53 function without increasing the amount of p53 protein. Alternatively, two novel possible mechanisms were demonstrated. First, core protein directly bound to the C-terminus of p53 and enhanced the amount of p53–DNA binding. The C-terminus of p53 protein is reported to regulate its DNA-binding affinity negatively [45]. Thus, binding of the core protein to p53 protein may mask the C-terminus of p53 and hence increase the p53–DNA binding affinity. The second possibility extrapolated from the results is the interaction between core protein and $hTAF_{II}28$. p53 interacts with TBP when exerting its transcriptional activity [46], and $hTAF_{II}28$ was reported to interact with TBP [47]. Because core protein itself did not possess transcriptional ability, these findings suggest that core protein may work as a transcriptional coactivator, possibly resulting in a complex of p53, core protein, $hTAF_{II}28$, and TBP that influences the transcriptional ability of p53.

HCV is known to be a causative agent of HCC [2]. It has been reported that the majority of HCC without a p53 mutation express p21 protein, whereas only a small fraction of HCC with p53 mutation demonstrate significant p21 expression [48]. p53 mutation was reported to be a poor prognostic indicator in patients with HCC [49]. Therefore, it is highly likely that once p53 is mutated, loss of the enhancement of p53 ability by core protein becomes apparent and the HCC may become progressive.

In conclusion, HCV core protein may directly promote cell proliferation and induce inflammatory reaction by activating SRE-, AP-1-, and NF-κB-associated pathways. On the other hand, core protein could enhance p53 function. These opposing functions may result in exquisitely balancing the proliferation of hepatocytes infected with HCV.

References

1. Choo QL, Kuo G, Weiner AJ, Overby LR, Bradley DW, Houghton M (1989) Isolation of a cDNA clone derived from a blood-borne non-A, non-B viral hepatitis genome. Science 244:359–362
2. Shiratori Y, Shiina S, Imamura M, Kato N, Kanai F, Okudaira T, Teratani T, Tohgo G, Toda N, Ohashi M, Ogura K, Niwa Y, Kawabe T, Omata M (1995) Characteristic difference of hepatocellular carcinoma between hepatitis B- and C-viral infection in Japan. Hepatology 22:1027–1033
3. World Health Organization (1997) Hepatitis C fact sheet. http://www.who.int/infs/en/fact164.html
4. Hijikata M, Kato N, Ootsuyama Y, Nakagawa M, Shimotohno K (1991) Gene mapping of the putative structural region of the hepatitis C virus genome by in vitro processing analysis. Proc Natl Acad Sci USA 88:5547–5551
5. Grakoui A, Wychowski C, Lin C, Feinstone SM, Rice CM (1993) Expression and identification of hepatitis C virus polyprotein cleavage products. J Virol 67:1385–1395
6. Treisman R (1996) Regulation of transcription by MAP kinase cascades. Curr Opin Cell Biol 8:205–215
7. Baeuerle PA, Henkel T (1994) Function and activation of NF-κB in the immune system. Annu Rev Immunol 12:141–179
8. Karin M, Liu ZG, Zandi E (1997) AP-1 function and regulation. Curr Opin Cell Biol 9:240–246
9. Levine AJ (1997) p53, the cellular gatekeeper for growth and division. Cell 88:323–331
10. Niwa H, Yamamura K, Miyazaki J (1991) Efficient selection for high-expression transfectants with a novel eukaryotic vector. Gene 108:193–199
11. Kato N, Yoshida H, Ono-Nita SK, Kato J, Goto T, Otsuka M, Lan KH, Matsushima K, Shiratori Y, Omata M (2000) Activation of intracellular signaling by hepatitis B and C viruses: C-viral core is the most potent signal inducer. Hepatology 32:405–412

12. Yoshida H, Kato N, Shiratori Y, Otsuka M, Maeda S, Kato J, Omata M (2001) Hepatitis C virus core protein activates nuclear factor κB-dependent signaling through tumor necrosis factor receptor-associated factor. J Biol Chem 276:16399–16405

13. Kern SE, Pietenpol JA, Thiagalingam S, Seymour A, Kinzler KW, Vogelstein B (1992) Oncogenic forms of p53 inhibit p53-regulated gene expression. Science 256: 827–830

14. Ishikawa Y, Mukaida N, Kuno K, Rice N, Okamoto S, Matsushima K (1995) Establishment of lipopolysaccharide-dependent nuclear factor κB activation in a cell-free system. J Biol Chem 270:4158–4164

15. Matsusaka T, Fujikawa K, Nishio Y, Mukaida N, Matsushima K, Kishimoto T, Akira S (1993) Transcription factors NF-IL-6 and NF-κB synergistically activate transcription of the inflammatory cytokines, interleukin 6 and interleukin 8. Proc Natl Acad Sci USA 90:10193–10197

16. el-Deiry WS, Tokino T, Velculescu VE, Levy DB, Parsons R, Trent JM, Lin D, Mercer WE, Kinzler KW, Vogelstein B (1993) WAF1, a potential mediator of p53 tumor suppression. Cell 75:817–825

17. Otsuka M, Kato N, Lan KH, Yoshida H, Kato J, Goto T, Shiratori Y, Omata M (2000) Hepatitis C virus core protein enhances p53 function through augmentation of DNA binding affinity and transcriptional ability. J Biol Chem 275:34122–34130

18. Regnier CH, Song HY, Gao X, Goeddel DV, Cao Z, Rothe M (1997) Identification and characterization of an IκB kinase. Cell 90:373–383

19. Schall TJ, Lewis M, Koller KJ, Lee A, Rice GC, Wong GH, Gatanaga T, Granger GA, Lentz R, Raab H (1990) Molecular cloning and expression of a receptor for human tumor necrosis factor. Cell 61:361–370

20. Rothe M, Sarma V, Dixit VM, Goeddel DV (1995) TRAF-2-mediated activation of NF-κB by TNF receptor 2 and CD40. Science 269:1424–1427

21. Cao Z, Xiong J, Takeuchi M, Kurama T, Goeddel DV (1996) TRAF6 is a signal transducer for interleukin-1. Nature 383:443–446

22. Schreiber E, Matthias P, Muller MM, Schaffner W (1989) Rapid detection of octamer binding proteins with mini-extracts, prepared from a small number of cells. Nucleic Acids Res 17:6419

23. Yin MJ, Yamamoto Y, Gaynor RB (1998) The anti-inflammatory agents aspirin and salicylate inhibit the activity of IκB kinase-β. Nature 396:77–80

24. Delhase M, Hayakawa M, Chen Y, Karin M (1999) Positive and negative regulation of IκB kinase activity through IKKβ subunit phosphorylation. Science 284: 309–313

25. Takeda K, Takeuchi O, Tsujimura T, Itami S, Adachi O, Kawai T, Sanjo H, Yoshikawa K, Terada N, Akira S (1999) Limb and skin abnormalitites in mice lacking IKKα. Science 284:313–316

26. Li Q, Antwerp DV, Mercurio F, Lee KF, Verma IM (1999) Severe liver degeneration in mice lacking the IκB kinase 2 gene. Science 284:321–325

27. Matroule J-Y, Bonizzi G, Morlière P, Paillous N, Santus R, Bours V, Piette J (1999) Pyropheophorbide-a methyl ester-mediated photosensitization activates transcription factor NF-κB through the interleukin-1 receptor-dependent signaling pathway. J Biol Chem 274:2988–3000

28. Shimoda K, Begum NA, Shibuta K, Mori M, Bonkovsky HL, Banner BF, Barnard GF (1998) Interleukin-8 and hIRH (SDF1-a/PBSF) mRNA expression and histological activity index in patients with chronic hepatitis C. Hepatology 28:108–115

29. Cacciarelli TV, Martinez OM, Gish RG, Villanueva JC, Krams SM (1996) Immunoregulatory cytokines in chronic hepatitis C virus infection: pre- and posttreatment with interferon alfa. Hepatology 24:6–9

30. Fukuda R, Ishimura N, Ishihara S, Chowdhury A, Morlyama N, Nogami C, Miyake T, Niigaki M, Tokuda A, Satoh S, Sakai S, Akagi S, Watanabe M, Fukumoto S (1996) Intra-

hepatic expression of pro-inflammatory cytokine mRNAs and interferon efficacy in chronic hepatitis C. Liver 16:390–399

31. Cerny A, Chisari FV (1999) Pathogenesis of chronic hepatitis C: immunological features of hepatic injury and viral persistence. Hepatology 30:595–601

32. Wang JT, Sheu JC, Lin JT, Wang TH, Chen DS (1992) Detection of replicative form of hepatitis C virus RNA in peripheral blood mononuclear cells. J Infect Dis 166: 1167–1169

33. Antewerp DJV, Martin SJ, Kafri T, Green DR, Verma IM (1996) Suppression of TNF-α induced apoptosis by NF-κB. Science 274:787–789

34. Beg AA, Baltimore D (1996) An essential role for NF-κB in preventing TNF-α-induced cell death. Science 274:782–784

35. Wang CY, Mayo MW, Baldwin Jr AS (1996) TNF- and cancer therapy-induced apoptosis: potentiation by inhibition of NF-κB. Science 274:784–787

36. Marusawa H, Hijikata M, Chiba T, Shimotohno K (1999) Hepatitis C virus core protein inhibits Fas- and tumor necrosis factor alpha-mediated apoptosis via NF-κB activation. J Virol 73:4713–4720

37. Ray RB, Meyer K, Steele R, Shrivastava A, Aggarwal BB, Ray R (1998) Inhibition of tumor necrosis factor (TNF-α)-mediated apoptosis by hepatitis C virus core protein. J Biol Chem 273:2256–2259

38. Robinson MJ, Cobb MH (1997) Mitogen-activated protein kinase pathways. Curr Opin Cell Biol 9:180–186

39. Ray RB, Lagging LM, Meyer K, Ray R (1996) Hepatitis C virus core protein cooperates with ras and transforms primary rat embryo fibroblasts to tumorigenic phenotype. J Virol 70:4438–4443

40. Moriya K, Fujie H, Shintani Y, Yotsuyanagi H, Tsutsumi T, Ishibashi K, Matsuura Y, Kimura S, Miyamura T, Koike K (1998) The core protein of hepatitis C virus induces hepatocellular carcinoma in transgenic mice. Nat Med 4:1065–1067

41. Roger JAG, Darerca O, Rookes SM, Gallimore PH (1996) Control of p53 expression of adenovirus 12 early region 1A and early region 1B 54K proteins. Virology 218:23–34

42. Chirillo P, Pagano S, Natoli G, Puri PL, Burgio VL, Balsano C, Levrero M (1997) The hepatitis B virus X gene induces p53-mediated programmed cell death. Proc Natl Acad Sci USA 94:8162–8167

43. Megyeri K, Berencsi K, Halazonetis TD, Prendergast GC, Gri G, Plotkin SA, Rovera G, Gonczol E (1999) Involvement of a p53-dependent pathway in rubella virus-induced apoptosis. Virology 259:74–84

44. Devireddy LR, Jones CJ (1999) Activation of caspases and p53 by bovine herpesvirus 1 infection results in programmed cell death and efficient virus release. J Virol 73: 3778–3788

45. Hupp TR, Meek DW, Midgley CA, Lane DP (1992) Regulation of the specific DNA binding function of p53. Cell 71:875–886

46. Farmer G, Colgan J, Nakatani Y, Manley JL, Prives C (1996) Functional interaction between p53, the TATA-binding protein (TBP), and TBP-associated factors in vivo. Mol Cell Biol 16:4295–4304

47. Mengus G, May M, Jacq X, Staub A, Tora L, Chambon P, Davidson I (1995) Cloning and characterization of hTAF$_{II}$18, hTAF$_{II}$20, and hTAF$_{II}$28: three subunits of the human transcription factor TFIID. EMBO J 14:1520–1531

48. Qin LF, Ng IO, Fan ST, Ng M (1998) p21/waf1, p53 and PCNA expression and p53 mutation status in hepatocellular carcinoma. Int J Cancer 79:424–428

49. Honda K, Sbisa E, Tullo A, Papeo PA, Saccone C, Poole S, Pignatelli M, Mitry RR, Ding S, Isla A, Davies A, Habib NA (1998) p53 mutation is a poor prognostic indicator for survival in patients with hepatocellular carcinoma undergoing surgical tumor ablation. Br J Cancer 77:776–782

Subject Index

The manufacturer's authorised representative in the EU is Springer
Nature Customer Service Centre GmbH, Europaplatz 3, 69115 Heidelberg,
Germany. If you have any concerns regarding our products, please
contact ProductSafety@springernature.com

Printed and bound by CPI Group (UK) Ltd, Croydon, CR0 4YY

28/04/2026

02098536-0004